The Company of Others:
Essays In Celebration of Paul Shepard

THE COMPANY OF OTHERS

Essays in Celebration of Paul Shepard

Edited by Max Oelschlaeger

KIVAKÍ PRESS

Durango, Colorado

Affirming the ties that connect our bodies, communities, and ecosystems.
Providing access to an ecological experience of life and place.

Kivakí Press, Inc.
585 East 31st Street
Durango, CO 81301

Kivakí Press offers special discounts for bulk purchases of books for promotions, premiums, non-profit or educational organizations. Arrangements can be made for subsidiary publication of special editions or book excerpts. Address inquiries to:

Kivakí Press Special Sales
585 East 31st Street
Durango, CO 81301
800-578-5904 tel • 970-385-1974 fax
kivaki@frontier.net (e-mail)

Preassigned Library of Congress Catalog Card Number: 95-60390

Publisher's Cataloging in Publication

The company of others: essays in celebration of Paul Shepard /
edited by Max Oelschlaeger.
p. cm.
Includes bibliographical references.
ISBN: 1-882308-24-7

1. Human ecology. 2. Shepard, Paul, 1925- I. Oelschlaeger, Max, ed.

GF8.C66 1995 304.2
QBI95-20221

Bear and elk illustrations by Olive Charles
Digital design by Olive Charles and Greg Cumberford
Text set in Optima

First Edition
First Printing 1995
Printed in the United States of America on recycled paper
1 3 5 7 9 10 8 6 4 2

TABLE OF CONTENTS

INTRODUCTION
Max Oelschlaeger

Overview

THIS VOLUME OF ESSAYS CELEBRATES Paul Shepard's life and contributions to the world of ideas, as a thinker cut out of a different cloth and as a teacher who deeply influenced the lives of many young women and men. No overview, of course, can do more than scratch the surface of his enormously rich and provocative life as a free-thinking scholar. Intellectual life in the twentieth century is increasingly fragmented, largely as a consequence of the dominant model of the university as a research institution and the concomitant proliferation of disciplines and knowledge. In consequence, the work of individuals such as Paul Shepard, thinkers who freely cross intellectual boundaries, falls through the cracks of the intellectual's mind. Even now Paul defies categorization for me; likely that is for the best. If I placed him anywhere, disciplinarily, it would be as a human ecologist. Yet that category itself is subject to great imprecision. Not that Shepard's work is imprecise.

Honesty requires a confession: I make no attempt to capture the full range of Shepard's ideas. Still, as this collection begins to make clear, a diverse array of serious thinkers has found much of merit, perhaps even "the cutting edge," in his work. Dave Foreman observes that Shepard is "the most important scholar of our time," suggesting that Nobel prizes are not, unfortunately given for the kind of "important scholarly" endeavor that Shepard represents.[1] Ironically, perhaps, the essays suggest that the significance of his ideas is far greater than most intellectuals, let alone popular readers, recognize. Because of the radical, even revolutionary nature of his writing, he is in some ways a man outside his own time. The greater part of Shepard's reputation likely lies ahead—in the next century where evolutionary thinkers are taken seriously—rather than behind. Which is to say, then, that future generations will come to appreciate his work more fully than the present, intellectually dominant, one. But the contributors hope that this collection marks the beginning of a more adequate consideration of Shepard's ideas. Of course, collectively considered, these essays only begin to contextualize Shepard's work in the larger currents of late-modern thought, especially evolution and interpretation. Even as the twentieth century draws near its close, evolutionary thought has yet to achieve cognitive hegemony—most Western intellectuals, let alone the masses, do not take time (in any of its myriad guises) seriously. Our political, sociological, economic and philosophical theorists still write as if an enormous intellectual revolution, perhaps beginning in the work of Charles Darwin in biology and Rudolf Clausius in physics, has not occurred. Accordingly, one way of characterizing Shepard's work, especially by way of distinguishing it from the work of contemporaries, is in terms of time: he takes time seriously. By taking time seriously Shepard escapes the specious present, which unthinkingly categorizes Man within an increasingly dogmatic framework of Enlightenment ideas, especially as these have been articulated through the discourse of the dominant political-economy. Such understandings of human being are, in contrast to Shepard's ideas, a thin gruel, indeed. If we are to escape the ideology of Man, Calvin Martin suggests, then we need "a cosmic lodestone that is miraculously . . . free of historicism, anthropologism, rationalism, relativism, positivism, structuralism, or any other cant, ideology, or agenda. . . . Paul Shepard, philosopher and population biologist, comes as near as anyone to furnishing such a universal reference point."[2]

What the Shepard oeuvre offers is a hermeneutic, a new interpretation of human being and culture, based on an evolutionarily informed "Book of Nature." What happens when we reread Western history against evolutionary texts? The upshot is a new story, a new narrative, a new conversation that includes the others and, ultimately, an alternative definition of human beingness, a definition that may undergird an age of ecology. Living as we do in a time of crisis, a crisis engendered by the insidious and relentless undercutting of the ecological systems that sustain life on earth, the work of a thinker like Shepard can only grow in importance. Each of the essays in *The Company of Others* gives reason why this is the case.

Shepard is one of a very few intellectuals who has the psychological wherewithal and the intellectual courage to take nature (evolutionary process) seriously. The consequence is a far-reaching inquiry that seriously considers the implication of evolution for our own self-conceptualization, our sociality, our intelligence, and our basic being-in-the-world in the company of plants and animals, that is, the others. From his first book, *Man in the Landscape: A Historic View of the Esthetics of Nature* (1967) to his most recently published essays, such as "A Post-Historic Primitivism" (1992), the evolutionary current runs deeply and powerfully.[3] His inquiry has been both deconstructive, as in *Nature and Madness,* which is a penetrating critique of the role of religion in fostering our sense of alienation from Earth, and reconstructive, as in *Thinking Animals,* where he assesses the implications of animals (through what might be termed an evolutionary psychology) in the development of human intelligence. It is fair to say, then, that Shepard offers a thick description of our human being, one that makes clear how much we have been shaped by the company of others. He persuasively fleshes out in concrete detail the meaning of the expression that the human species is nature grown self-conscious.

Personal Reflections on the Man and His Work

My own interactions with Paul mirror, in one sense, his situation more generally in intellectual culture. While I find him one of the most provocative thinkers for anyone attempting to do environmental philosophy, one only occasionally finds him mentioned in the field's scholarly journals, and most often he is being attacked for his views on hunting. Yet, even once one has learned of his work, it is not easy to lay one's hands on it. Only large university libraries have his books, and even then the collection is often incomplete. *The Subversive Science* is his best known book, but this is an edited collection, containing only a short essay written by himself, which offers at best an elliptical suggestion or two as to the richness of his overall work. And the publishers, of course, have done a poor job in keeping his works in print.

My own case history in coming to know his work is, I think, typical, in the sense that I learned of Shepard adventitiously rather than by design: I came to know him only because of George Sessions' tireless effort to identify scholars whose ideas were relevant to deep ecology. (See Sessions' contribution in Part II.) Virtually all of us have more reading to do than we can possibly accomplish; only at Sessions' insistence did I start reading Shepard's books. Yet, these difficulties aside, having read Paul's published books, the majority of his essays, and his major work (now titled "The Others" in manuscript) on the relation of humans and animals, I can say that even though his influence on me has already been substantial, what lies ahead will surely be of greater consequence. For Paul challenges all of us, and this is crucial for environmental philosophers, to call the category of History into question.

Philosophers, of course, entertain themselves by making outrageous claims about "critical thinking." Yet the history of philosophy has been based on dogmatic and now sacred assumptions that human beings are actually only disembodied minds, that the most consequential aspect of minds are arguments that obey the three laws of thought, and that the natural world is a realm of mere experience, transitory appearances that mask underlying conceptual realities. Shepard, to put it mildly, awoke me from my dogmatic slumber. By calling history, including the history of philosophy, into question, the legitimating narratives of Western culture, the stories that tell us who we are, what is of value and meaning, what is good and what is evil, and so on, are themselves called into question. Such a bracketing of history, in some ways like the phenomenologist's epoche (yet in other ways unlike), is a radical move, for it frees our thinking from the "hardening of the categories" that characterizes the modern age. For example, the story that names us Homo sapiens, the rational animal, looks quite different through the lens of the deep past. We begin to see that Plato and Descartes, and all the other philosophers who have participated in legitimating the story that we are disembodied thinking things, made their claims by not only ignoring but indeed obliterating any account of the influence of evolutionary process upon human intelligence.[4] Or, to take another example, the story that tells us that beauty is the invention of aestheticians, and that it is not knowable apart from aesthetic theory, is itself countermanded by an evolutionary narrative that begins, if in no other place, with the eye.[5]

More recently, Shepard's work on the others, especially the animals that have helped to shape our basic being, has been weaving itself into my thinking. And my practical life as well. I find myself, once a philosopher who wrote papers on the excluded middle, now writing on raptors and their flights through the human imagination. (Interesting, isn't it, that Paul himself was once a falconer? Did his own intellectual life take wing with these magnificent creatures, good, wild, and free?) I find myself reconsidering Bear—a theme found in several essays herein. I suppose, finally, that Paul bears (sorry) responsibility for the relocation of the family household to New Mexico, for he brought to my attention the literal need for wild environs, for excursions into the bosque along the Rio, into the alpine forests of the Sangres, and walks along gravelly Sandia trails. I am, admittedly, a long way from being a reinhabitant of the desert southwest: no pretense there. But Paul insists that culture, and by implication person, is a mosaic, one that can be reconstituted through recovery of pieces of the wild.

Finally, I have learned from my friend, Paul Shepard, that what renews us as humans are the same energies that course through the great round of being, the marvels and mysteries, indeed, the beauty of the good green Earth, and the company of others.

The Essays

The contributors' works are grouped into three parts. They speak for themselves. Given the length of the festschrift, I shall forego detailed commentary on the essays. Part I, The Man, His Work, and Influence, contains essays that are of a more biographical and personal nature than those in Parts II and III. Flo Krall provides a revealing biographical study of Paul, and other essayists provide details of Shepard as a dynamic teacher and as an iconoclastic scholar. Part II, Reflections on Shepard's Ideas, offers essays that assess his influence and explore some of the many aspects of his thinking. Finally, Part III, Essays in Honor of Paul Shepard, contains essays that bear obliquely on Shepard's ideas, but were written for this festschrift by friends who wished to honor him.

Notes

1. Dave Foreman, *Books of the Big Outside* (Tucson: Ned Ludd Books, 1994), 25.
2. Calvin Martin, *In the Spirit of the Earth: Rethinking History and Time* (Baltimore: Johns Hopkins University Press, 1992), 139.
3. Appendix 1 offers a relatively complete listing of Shepard's works and publication history.
4. On this point see *Thinking Animals* and *The Sacred Paw.*
5. See *Man in the Landscape.*

PART I

THE MAN, HIS WORK, AND INFLUENCE

The Old Bones: For Paul Shepard

Gary Snyder

Out there walking round, looking out for food,
A rootstalk, a birdcall, a seed that you can crack
Plucking, digging, snaring, snagging,
Somehow getting by.
No good out there on dusty slopes of scree—
Carry some—look for some.
Go for a hungry dream.
Deer bone, Dall sheep,
Bones hunger home.
Out there somewhere
A shrine for the old bones,
The dust of the old bones,
Old songs and tales.
What we ate—who ate what—
How we all prevailed.

Riding the Thermals: A Biographical Sketch of Paul Shepard

Florence R. Krall

> All art should have some more mystery and meaning to it than is apparent to a quick observer.
>
> —Henry Moore[1]

LONG BEFORE I MET PAUL SHEPARD, I read his books. My students and I pondered their meaning, not really understanding why but somehow being transformed by the language. His writing, and the thinking it reflected, brought a speculative tone, rather than theories all set out clearly for us to take and apply—one that proposed the ambiguous and complex business of being "thinking animals." "What kind of a man would think of such things?" we would ask ourselves.

Now as his partner—and yet not able to answer that original question—I have expanded it to another: "What experiences of life led this man to be who he is?" As a teacher, I have always been interested in the significant motivations that people encounter that open them to their potentials. Thus my main purpose in this biographical essay will be to uncover bits of Paul Shepard's past that reveal significant moments in the course of his life.

I begin with our journey to his fiftieth high school reunion in Mountain Grove in the Ozarks of southern Missouri. On our way, we stopped in Kansas City, Missouri. Paul had reserved a room in the section of town where he entered the world. From our window we could see St. Luke's Hospital where he was born, with the help of Dr. Brutus Hamilton, to Clara Louise Grigsby Shepard and Paul Howe Shepard on July 12, 1925.

In a shady, quiet neighborhood on Fifty-ninth and Kenwood, we visited the home where he, his mother and father, his brother Dick (two years his junior), Mary Jones, their housekeeper, and Punch, the family dog, lived until 1934. The house is a frame bungalow with a porch with lattice trim and a swing, bay windows in the dining and living areas, and a small bedroom upstairs where he and his brother slept. The watermelon tank his father had set into the back yard for swimming is gone, but the house has not changed much, Paul says, except that then it was painted white instead of the present gray.

We walked around the neighborhood where, at age seven, he delivered Liberty Magazine, a weekly fiction magazine, to a dozen clients. Not far away is a small mall that was once a Piggly Wiggly Grocery Store where he bought Bond Bread at a nickel a loaf for his mother. Nearby he showed me the sidewalk curb where on most days on the way home from school he sat to eat the remaining sandwich in his lunch box, probably to avoid the reprimands of his mother who, on constant vigil over her sons' health, insisted on a good diet, exercises each night to combat flat-footedness, and avoidance of cold drafts after playing hard and sweating.

A block from his home, Paul stood on the street corner where he recalls finding a box turtle one day. When he took it home, he was cautioned by his grandmother not to put it in the aquarium which his father had installed in the dining room to house assorted fish, turtles and pollywogs he brought back from fishing trips. This was a "terrapin," Grandmother Grigsby informed him, a land dweller. That there were different kinds of turtles with different habitat requirements impressed "Buddy" Shepard. At this early age he was already developing a keen interest in animals, their differences and uniqueness, as is evident in a letter to the Kansas City Star in 1933, his first attempt at journalism.

> HAS ANYONE SEEN PUNCH? Buddy Shepard is Ill in Bed, So Can't Hunt for His Dog. Dear Star: I have lost my dog. He has lived with us seven years. Do you think dogs can smell as well in snow in winter as they can in summer? He is black and white and has a little brown on his face, and has a tail about three inches long. He has a big black spot on his back that looks like a saddle. He is about a foot and a half high. He will come by the name of Punch. I am sick in bed and can not go out and look for him. I am 8 years old. Please help me find him. He is a wire-haired terrier. Buddy Shepard, 5925 Kenwood.

In this first publication, we see cognitive tendencies that would develop later in manhood and characterize his life work: heuristic thinking and acute observation, decisiveness and ability to act in the face of adversity, an underlying faith in rationality, and fascination with animals and their characteristics. Punch, by the way, was returned home safely.

We next visited Troost Elementary School, built in 1922 of light brick, rock and concrete which is relatively unchanged except, he says, for the bars on the windows on the ground floor. Our steps resounded over wooden floors as we worked our way past janitors busy with summer cleaning to the stairs that led to his kindergarten room where he and classmates sang "Good Morning" to their teacher each day and napped on their blankets in the afternoon. His other memories of these school days include the tactile sensation of creating and painting a small clay dish and the "tangible sense of pleasure" brought by the vivid colors of his crayons. We worked our way back to Kenwood past the church on the corner where on the lawn, now a parking lot, he and his best friends, Donald Cushing, Bill Wagner and Bill (Zimmie) Zimmerman played boy games and flew kites. His heroes in those years were Charles Lindberg and Tom Mix, the film cowboy. Our next stop was at the Nelson-Atkins Museum of Art with its Henry Moore Sculpture Garden. I had been introduced to Henry Moore's sculptures and Paul Shepard in the same year. It seemed appropriate that on our way to the fiftieth reunion of Paul's high school graduation class, we should stop here in his birthplace where I could pay homage to that other artist who had enriched my life. I had the garden to myself. The bronze statues glistened in a fine drizzle as I walked from one to the next astonished by their simple beauty. Both Paul Shepard and Henry Moore seem to me to have been honed of the same vision and dedication to their art, which upon closer examination, brings multi-faceted meaning.

After wandering to my heart's content, I met Paul in the foyer and we walked to the Country Club Plaza. In a small cafe over lunch, he related how he and his friends would come there for movies and treats. Designed in Santa Fe architectural style, it was one of the first planned shopping centers in the United States. It is preserved much as it was first built in 1925. His grandmother, Anna Marie Grigsby, widowed in 1930, lived nearby at the Alcazar Hotel. He remembers his visits to her apartment—the lobby with its great chandelier and the candy jar of orange slices always available.

We left the Plaza, packed our car, visited the graves of his parents and grandparents, and then followed the same route that he and his family had taken when they moved to Mountain Grove. It was February in 1934, when they along with Grandmother Grigsby, Mary Jones and Punch packed themselves into a Nash sedan and headed for a new life, one that would be filled with fond childhood memories for Paul.

The last years had been hard ones for the senior Paul Shepard. The Depression had taken its toll and obliterated over a decade of his work as an orchard grower. It took his own Select Farm as well as the house on Kenwood. Having graduated in horticulture at the University of Missouri and established a name for himself in the Kansas City area for his successful fruit growing, he was appointed director of the Missouri State Fruit Experiment Station in Mountain Grove by Governor Lloyd Stark.

The Fruit Station had been established by the Missouri Assembly at the turn of the century. Nestled among giant oak, hickory and tulip trees, the house, the beautiful old Tudor administration building, and adjoining structures covered ten acres on top of a hill over-looking almost two hundred acres of experimental orchards, the surrounding farms and woods, and the town of Mountain Grove. Here the Shepards found a home that they would cherish—a hospitable place open to all of their friends, a place where boyhood would be firmly grounded for their sons. In talking to him and friends about his formative years, I see on the surface two significant influences, place and parents, especially his mother, Clara. Place and parents complemented each other creating a supportive environment for a young boy who seems to have been filled with intense interests and a creative spirit. His father, a typical head of the household in those years, was a serious man with passionate dedication to his work, the heir of a Yankee Puritan outlook. As was expected of fathers of his day, he provided the means for a good life for his family, shared hunting, fishing and athletic forays with his sons, but was not himself highly involved in their day to day nurturing. He was, however, behind the scenes, always keenly attentive, checking on his sons' progress and suggesting alternatives for them. The primary care was left to Clara, who also, in the style of the times, was the traditional lady of the household. She was freed from most of the menial tasks about the house by Mary Jones, the daughter of a slave, the only Black housekeeper Mountain Grove had ever seen, and a member of the Shepard household for forty years. Clara was able to devote her energies to children and friends and her own inclinations.

A friend of Clara's during this period, perk and pretty Susan Olds Jones, now in her nineties, speaks of Clara's filial devotion and character. "The minute you met Clara Shepard," she says, "you knew she was somebody." According to Susan, Clara had keen interests of her own as well as a sincere interest in others and maintained an active and highly involved community and family life: She painted, played the piano, acted in local dramas, belonged to a bridge club, and entertained friends whom she cherished throughout the years. She was playful, filled with humor, and somewhat of an extrovert, dancing the cancan on occasion, speaking French, playing the piano, and taking off her bra on trips to the city with her women friends so as not to be bound down. She loved Punch and his mate, Judy, who joined the family when they moved to Mountain Grove and brought forth a litter of pups which Clara took with her to her bridge club in a huge basket tied with a bow. However, her first devotion was to her sons and their projects, which she and the senior Paul supported unconditionally, says Susan. Sensing that Clara may have been an overindulgent and overprotective mother, I asked Susan if she were dedicated solely to her children. Ann Grey, Susan's daughter interjected without hesitation that Clara mentored and showed a special interest in her as well as other people in the community. For example, she took food to a woman who had become alcoholic and, during World War II, mourned the loss of her nephew as if he were her own child.

Paul remembers those years as ones filled with unrestrained exploration of the natural world. In *Nature and Madness*[2] he says,

> [A child] will learn that his childhood experiences, though a comfort and joy, were a special language....Thenceforth natural things are not only themselves but a speaking. He will not put his delight in the sky and the earth behind him as a childish and irrelevant thing.... He will not graduate from that world but into its significance. So, with the end of childhood, he begins a lifelong study, a reciprocity with the natural world in which its depths are as endless as his own creative thoughts.[3]

When he wrote these words, Paul was, I believe, speaking of himself and of his childhood.

Growing up close to the natural environment where one's needs and interests are taken seriously yet are unrestricted, brings self-knowledge that borders on both selflessness and self-possession, but leads to a psychological orientation quite different from empathy or altruism. One situates himself as a part of the surround, not as a self-centered child that sees everything as an extension of himself, bound to others in need of love and devotion, but as separate and independent, yet joined inextricably to the natural world and its creatures. Like the treasures of the natural world, ideas and opportunities would appear before Paul and provide sustenance for an intellectual life that felt its way rationally through a web of earthly connections. These ideas would be reworked through the vagaries of maturing, over and over, as when Henry Moore reinterpreted the reclining figure and its relationship to the primary feminine. For Paul Shepard, the theme that developed early in his life, and that he continues to refine, is our indebtedness to animals and our paleolithic past for understanding and living an appropriate human ecology.

As a young man Paul would search unsuccessfully for the complete feminine he knew in the idyllic days of childhood as biological mother and nature. He would become a father and scholar unconditionally devoted to his children and to those "Others," the creatures that without any demand of recompense had filled his life with adventures and joy. He would accept his origins in the feminine that could be nurturing as well as destructive and was not meant to retain the grown man in womb-like comfort. At a deep level he would become ever more saddened by the exploitation of the Earth, yet, in the face of disappointing human interactions and ecological problems, he would remain hopeful, spirited and witty.

Clara and experience delivered him into the ambiguity of the real world. Although I stress Clara's profound influence on her son, the course of his life was not dependent on her. A mother is not solely responsible for the child's development, for it is often the impossibility of any human, mother or father, to meet completely a child's needs that propels that child toward self-assertion and motivation. Complementarity and reciprocity of feeling between child and mother may be as much a matter of genes and chance as are strained relationships and tension. Although I believe that Clara Shepard was a significant figure in Paul's ontogeny, planned and unforeseen encounters also changed his life. From his beginning he seems to have had an uncanny capacity to turn adversity into good luck. He also was able to select intellectual substitutes for the nurturing mother who would act as his mentors and guides.

Four periods emerge from Paul's life thus far, each comprising approximately 20 years. The first includes his childhood, youth, service in World War II, and undergraduate studies. This was a gestation period, a time of exploration, of trying on roles and of passing from one intense interest to the next. The second includes his young adulthood, when he married and fathered three children, completed his graduate degrees at Yale University, and labored over his first book which contained the outlines of his future work. During this period he was filled with an appetite for learning and experience; it ended in the sixties where in the midst of social upheaval he found himself in a new place physically and psychologically. Fledged to manhood, he sought unexplored country, new ways of seeing and acting, alternative teaching positions, and a new marriage. This third period, during which he published many of his books and essays, was the most prolific period of his life, although one of unease until he found his niche in an endowed chair at Pitzer College and the Claremont Graduate School. The fourth and most recent phase of his life, as an elder scholar and grandfather, is just commencing. It carries with it a more comprehensive and complex view of the world. His wit remains intact as does his exuberance over new discoveries. Paul retains the capacity for bursts of creativity and action driven by a sense of timing and ethics. Buttressed with years of assiduous research that crosses the fields of science, history, and anthropology, his voice in ecological circles is strong and critical, buffered with asperity that acknowledges the contradictory aspects of life. The force of his ideas and language is contradicted by a gentle demeanor and mindful respect of other positions.

His present orientation is exemplified by his response to my daughter, who asked his opinion about friends who were experiencing a religious revival. He replied that he has no problem with spiritual experiences, whatever form they take. He himself is pantheistic and feels that spirit resides in all things and reveals itself in diverse ways. Everything in nature is numinous, he said, and it does not surprise him at all that people have religious awakenings. An aside in closing left my daughter with her mouth agape. "I had not thought of such things," he said, "until at home after burying my mother, I found she had returned as a Canyon Wren."

~ ~ ~

I return to Mountain Grove and his fiftieth high school reunion. The square at the center of Mountain Grove was deserted, except for myself and someone taking a nap in the bandstand where Paul played the flute in high school band concerts on hot summer evenings. His class was gathered in the Sill Center a few blocks away waiting to have their reunion picture taken. The break gave me time to think through friends and festivities and to give more attention to the square which in the morning was filled with returnees greeting each other in astonished recollection.

I walked around the square and paused before the memorials to "American Soldiers who died in defence of the Flag." Mountain Grove from its first days has been a conservative, white enclave. As a requirement for incorporation, the square was established and city officials elected in 1881. Earlier "a general store and post office called Hickory Springs" sprang up near a spring and a grove of hickory trees, where Alumni Park stands today, to trade with hunting parties of Osage Indians. French and Spanish settlers, mining lead and zinc, were followed by "early pioneers from Tennessee, Virginia and Kentucky" who were trappers, and hunters, and traders. They grew a little corn and tobacco but found the forests provided for the rest of their needs.[4] That morning the square had been packed with alumni, some descendants of those early settlers. The parade progressed around the square, with alumni walking or riding on hay racks, convertibles, or trucks as they celebrated fiftieth and twenty-fifth reunions. They were joined by Shriners on golf carts, a mounted posse, and a clown and his family riding monocycles. Following the parade, everyone mingled in the square—a time for recognizing each other and discovering what had transpired during the intervening years. The crowd finally dispersed into individual class picnics or lunches and would reunite once more that evening at a communal dinner at the Middle School.

From the square, I could hear an approaching train. The whistle sounded mournfully as the train rumbled through town and past the grain elevators only a few blocks away. One of Paul's early passions was for trains that, in those days, dramatically spewed steam and smoke as they ground to a stop. Their schedule was regular, and, when he heard the trains approaching, he would ride his bicycle to the railway station. He was friendly with engineers, took pictures and compiled a childhood "ethnography" on the trains and their crews. I can imagine him watching them puffing out of the station, looking beyond them down the track visioning the lands beyond.

His keenness for travel in later years grew from family vacations. Paul's first memory of such a vacation occurred the summer he turned three. They traveled by train to Florida with his mother's college friend, Marian Barker, and her son, Bobby, who was Paul's age. He remembers the train, the beach where he played in the sand, and the glass-bottomed boat. Relatives were visited in Chicago and Santa Fe, where in 1938, Clara arranged a visit between Paul and author and artist Ernest Thompson Seton, at his boys' camp, and whose adventure books, such as *The Two Little Savages*, Paul had read avidly. Clara arranged for Paul, then fourteen, to accompany the Future Farmers of America club, although he was not a member, to the New York World's Fair. The next year the family traveled to the San Francisco World Fair.

Like his father, who was a football star at the University of Missouri, Paul loved athletics, basketball and track, but especially football. He was a quarterback in high school, and didn't do badly, although he never equalled his father. To this day, he loves a good game with family members. He delights in watching his grandchildren participate in sports as well as rerunning plays with his former teammates.

Paul has no memories of being contained or restricted during his formative years; he ran freely through the countryside as a young child. With the help of a net his mother fashioned from cheese cloth, he began a butterfly collection, later followed by a bird egg collection—one that brought him lasting pleasurable memories. In his forthcoming book, *The Others: Animals and Human Being,*[5] he describes his affinity for these eggs.

> Their colors gave me at ten the palpable pleasure that crayons had given when I was five. The exquisite markings on the shells, different in each species and each individual, were messages in unknown writing...every egg of a different species was a summons and a wonder....Memories of boyhood collecting somehow inform...my theory that the human species emerged enacting, dreaming and 'thinking' animals, and cannot be fully itself without them.[6]

He grew up roaming first around the Fruit Station and then through the oak-hickory forests, farms, and brushy fields in landscapes which had not been altered significantly since settlement. He swam in and paddled on the ponds developed by his father immersed in throbbing choruses of frogs and stridulating insects. He fished and hunted with his father, and followed friends, especially the Rook boys whose father worked at the Fruit Station, on their rabbit and muskrat trapline. He admired them and their knowledge of the woods, and sometimes they took him hound-hunting at night. There is a photograph showing them proudly dangling a dead fox.

From childhood on, he lived in the company of animals. The family dogs were not his only pets; he kept owls and crows in the barn, box turtles and opossums. He received a raccoon one Christmas; it promptly bit his finger, but remained in the family for many years.

During middle school years he attend a summer camp, Camp Gravois at Versailles, Missouri, and was involved in crafts and nature studies. But it was not enough that Paul should study nature. Clara saw to it that he and his brother took lessons in art, dance, and piano. Later in high school he played the flute and was active in dramatics.

In adolescence Paul would take up falconry and continue this obsession for the next decade. His interest was first piqued by articles by the Craighead brothers written for *National Geographic*. Later, attending a University of Missouri football game, he saw Bob Hecht, a graduate student at the university, fly his Peregrine Falcon at halftime. The next summer he visited Hecht at Big Spring State Park, where he was a seasonal park naturalist, to talk about hawks. Soon he began training his own birds, which through the years would include Red-tailed Hawks, a Cooper's Hawk, and several Kestrels.

When I asked his classmates what they most remembered about Paul, they most frequently mentioned his intense love and devotion to hawks. Lester Barker remembers that after a call to the principal from Clara, he and Paul were excused from school to search for Napoleon, a Red-tailed Hawk who had escaped. Lester drove Paul's 1931 Model A Ford down country lanes with Paul standing on the fender, swinging a piece of meat from a cord and whistling for Napolean who finally returned to him.

Much of the training of these raptors took Paul into the countryside alone, but according to Susan Olds Jones, he was never "a loner." Raising, training and following his raptors influenced him significantly and brought him to that sacred ritual, the taking of one life to sustain another. Raptors remain the icons of life for him. One day, seeing a "red-tail" riding the thermals, he remarked, "What an ideal life! Perfectly in tune with the environment, they follow the wind currents and sun to plentiful food and optimal conditions. Everything they need lies before them."

Early on, Clara seems to have been guiding Paul toward a career in journalism. When he was nine, she helped him produce a newspaper, The Weekly Chit Chat, that he delivered to the families at the Fruit Station with a cart drawn by Judy, their wire-hair terrier. As a junior in high school, he served as editor of the school paper, *The Panther News*. His editorials, coming in the midst of World War II, reflect the patriotism hewn in stone and bronze in the Mountain Grove Square as well as family stories told to him of ancestors who fought in the Revolutionary and

Civil Wars. Sitting behind the typewriter, questioning the authority of administrators and mindless actions of students, he became the conscience of the school:

> Hey! We're still at War:...CERTAINLY STUDENTS need only a reminder—Stamp sales are dropping. The smug overconfidence that accompanies the I-bought-one-once people is what can beat us from inside. Again tomorrow morning stamps will be sold during first period classes. Tie that string on your finger now and "Buy a stamp and lick the other side." Feb. 14, 1942.

After one year he was "fired" as editor, he says, because he and his cohorts would gather on Saturday mornings and produce the paper without involving the journalism class or teacher. During the summer following his junior year, however, with Clara's encouragement, he attended a Northwestern University journalism workshop for high school students, a challenging and memorable experience for a young man with a bent for writing and independent thinking. The experience, however, deterred him from a career in journalism; his impression upon leaving was that "journalists didn't know anything.... They were committed to reporting or writing opinions, but they were not trained in anything except how to write."

Patriotism and war were astir in him. At his father's suggestion he took the tests and qualified for the ASTP (Army Specialized Training Program) and was the first in his graduating class to enlist in the Army. Following graduation, he attended the University of Missouri as he awaited induction impatiently; in six weeks he dropped his classes and volunteered for service. A letter in his files, a reply to his father's inquiry about the status of credits Paul had received that summer, indicates the extent to which the senior Paul, behind the scenes, was ever watchful of the course of his son's life.

Paul was sent to Camp Roberts near Paso Robles, California for basic training as a gunner on 105 mm howitzers. From there he went to Santa Rosa Junior College, a clearing station for ASTP, where he met Mal Wood, who would become a life-long friend. They were transferred to the New Mexico College of Agriculture and Mechanic Arts at Las Cruces, New Mexico in January of 1944 where they completed a quarter of college training whereupon the ASTP Program was dissolved, and in March Paul was sent to the Twelfth Armored Division at Camp Barkley, Texas where he served as a radioman. In October, 1944, the division was shipped to Tidworth Barracks on Windmill Hill in England and in November went ashore at Normandy and crossed France to take part in the Battles of the Little Bulge in the Rhineland and central Europe.

With the ending of the war in Europe, he was assigned as an information and education specialist to the Army of Occupation in Heidenheim, Germany. For six months, he edited, engaged a publisher, and supervised the printing of his first book, a history of his battalion in combat: *The Pictorial History of the 493 Armored Field Artillery Battalion, 1943-1946.*[7] He was then given the option of attending college and he chose the University of Neuchatel, in Switzerland, where he studied German, geology, Swiss culture, and parasitology taught by Jean von

Baer. Paul would carry from that time a strong impression of this dedicated parasitologist who shocked his students by announcing that he himself was host for tapeworms studied in the class.

Paul regards the Army in the same way he does various experiences that may have not have been ideal. He insists he enjoyed basic training where daily necessities were provided, physical training was demanding, and expectations were clear and rewarded. Although options were limited and activities confined to others' schedules and demands, within these parameters, he was freed to hone his skills, develop friendships, and read. Perhaps now looking back, he idealizes a situation that was dramatically different from how he sees life in the years since, when, he says, "existence constantly seems to fall away at the edges."

Unlike basic training, war is not an easy place. Even in the artillery, spared of close combat, one must endure cold, sleep deprivation, discomfort and fear. His unit liberated the German concentration camp at Landsberg, where torched, emaciated bodies stacked like cordwood, and only three remaining live prisoners, starved nearly to death, are not easily forgotten. But there were moments of levity. On one occasion while deer hunting, he encountered a squad of German soldiers cut off from their army and ready to surrender; he brought them back as prisoners.

Reading through the letters Paul wrote home during those years, one can trace the growth and maturation of a naive, impressionable youth to a young man, matured by the experience of war. With more time on his hands in the Army of Occupation, he read insatiably. He discusses the authors in his letters, among them Shakespeare, who he feels was an explicit intellectual and literary influence. In answer to his mother's question about future plans, in one letter he says that although no progress had been made in politics since Washington's time, he was considering law, economics and politics. "On the other hand," he added as an afterthought, "there is a natural world out there I don't want to miss."

Paul returned to the United States on April 15, 1946, and was discharged at Jefferson Barracks, Missouri. After a summer of working at the Fruit Station, attending weddings, and visiting friends, he and his newly caught Cooper's Hawk, Clarabell, took up residence in Columbia, Missouri where he enrolled in the University of Missouri in the fall of 1946. He chose majors in English and wildlife conservation.

The program in wildlife conservation was situated within the Cooperative Wildlife Unit, directed by Rudolph Bennitt, a zoologist; the program was the first of its kind, bringing together the University, the State Conservation Commission, the Wildlife Management Institute, and the United States Fish and Wildlife Service. Bennitt became Paul's major professor, adviser, and mentor. (They had first met at the Fruit Station in 1936 when Bennitt visited concerning matters involved with his Bobwhite Quail population study. Paul, then a young boy, was impressed with Bennitt's interest in his bird egg collection and help in correcting Paul's wrong labels.) It was an exciting time when he found himself at conferences in the company of Aldo Leopold, the author of *A Sand County Almanac.* Independent studies with Bennitt on the essays of naturalist/teacher William Morton Wheeler and Gordon Walls's *The Vertebrate Eye* were major influences in his developing ecological view.

As would be a pattern throughout his life, he sought experiences outside the set academic program. In the summer of 1947, Paul enrolled in ornithology courses at Cornell University taught by Arthur A. Allen, who remained a lifelong friend. Subsequently, he was a teaching assistant in ornithology at the University of Missouri where he and Dick Youse, a fellow student in Wildlife Conservation, honed their skills in bird watching each morning. This intensive immersion in ornithology fused remembered bird-song from youthful bird-egg collecting days with names and behavior. Bird identification and watching would become a lifetime joy; above all other sounds—human talk, city noise, music—bird song would always come to him clear and sweet. In the summer of 1948 Paul was appointed seasonal naturalist at Big Spring State Park, where five generations of his family had camped and picnicked. While there, he became aware of logging operations within the park and wrote a letter to the governor protesting the policy. This was the beginning of environmental activism that would escalate in the years to come.

Upon graduation from the University of Missouri in 1949, he spent the summer as a seasonal naturalist at Crater Lake National Park, and applied for work with the Conservation Federation of Missouri. In the fall he became field secretary of that organization, then directed by Charles Callison, and was responsible for organizing conservation clubs for sportsmen. Callison asked Paul if he wanted to go public with the logging problem he had uncovered at Big Spring, and together they brought the issue before the public.

It was a cold February day when Paul, in an encounter with animals that would leave a lasting impression for the rest of his life, abruptly entered a new phase of his life. He was driving his new 1950 red Ford convertible to a conservation organizational meeting at Maryville, when he lost control on an icy stretch of highway and crashed into a truckload of pigs. The accident put him in the hospital for six weeks with a badly broken leg. But as was his bent, he turned adversity into opportunity. While in the hospital, he courted his nurse, Melba Wheatcroft, and, at Clara's suggestion, applied to a newly organized masters degree program in conservation at Yale University. August found him and his bride setting up their household in student housing in New Haven, Connecticut where Melba would find work in a hospital and he would complete masters and Ph.D. degrees.

The conservation program at Yale was directed by Paul Sears and G. Evelyn Hutchinson. The first seminar conducted by Hutchinson planted the seeds of interdisciplinary studies. Sears was also grounded in many disciplines, including geology, soils and plant ecology. The new graduate student was appointed as laboratory assistant to Sears, and he sandwiched additional experiences into summers and Christmas vacations. In the summer of 1951 he was a seasonal park naturalist at Glacier National Park; in December of that year, under the auspices of the American Museum of Natural History, he conducted a field study of the Bermuda Cahow at the Bermuda Research Station. He graduated in 1952 with an

M.S. degree in conservation. His final paper, entitled simply "A Study in Landscape Interpretation," an ecological study of the perception of the landscape by painters of the Hudson River School, identified attitudes implicit in their paintings. Paul later traveled to the original sites documenting the ecological changes with photographs. That spring a son, Kenton Howe Shepard, was born to the Shepards, and in the summer the family returned to Paul's work at Glacier National Park. In the fall he began a doctoral program, an interdisciplinary degree combining work in ecology, landscape architecture, and the history of art. Paul Sears (conservation) chaired his committee and members included Ralph Henry Gabriel (American history), G. Evelyn Hutchinson (zoology), William Jordy (art history), and Christopher Tunnard (landscape architecture).

This was a busy time for Paul as he assumed some of the child care and also became highly involved nationally in the environmental movement. He had been granted the Arthur Hoyt Scott Garden Award by the National Council of State Garden Clubs, with a membership of 250,000, and in 1953 was invited to be their national conservation chairman. As such he became a member of the Natural Resources Council of America and took active part in lobbying in Washington with such people as Dave Brower (Sierra Club), Charles Callison (National Wildlife Federation), Ira Gabrielson (Wildlife Management Institute), Joe Penfold (Izaak Walton League), Fred Packard (National Parks Association), and Olus Murie, Sigurd Olson, and Howard Zanhniser (Wilderness Society). He and Rachel Carson, with their common interest in the effects of pesticides, also crossed tracks at testimonies in Washington.

In 1954, he received his Ph. D. from Yale University; his dissertation topic was entitled "American Attitudes Towards the Landscape in New England and the West, 1830-1870." Upon graduation he accepted a position as professor of biology and director of Green Oaks, the college field station, at Knox College in Galesburg, Illinois. The position was especially inviting because of the 850 acres at Green Oaks where he and students would work to establish a tall grass prairie and develop an all-college program for use of the area. The following year in 1955, a daughter, Margaret Elizabeth Shepard, was born. His final summer of naturalist work was at Olympic National Park in 1956. After the season's work at the park ended, he became a whistle-blower once again, as he had at Big Spring State Park. In Olympic National Park the problem was more serious. With the benefits of his research on attitudes toward the landscape, continued participation in critical environmental issues at a national level, his grooming under the tutelage of Charlie Callison, and a capacity for quick action and eloquent language, Paul was an ideal spokesman for environmental causes. Furthermore, his ecological training had helped him to see the forest as well as the trees.

Carsten Lien in his history of the scandal at Olympic Park, *Olympic Battleground: The Power Politics of Timber Preservation*,[8] recalls how the seasonal naturalist staff, lead by Grant Sharpe, concerned with indiscriminate logging in the park, had been holding "logging the park" sessions concerning these operations. In the summer of 1956, Paul replaced Sharpe (who was completing his dissertation) as seasonal naturalist. According to Lien,

> Shepard, from the moment of his arrival, was stunned by the ever-present logging occurring everywhere in the park. His reaction proved to be enormously validating for the rest of the seasonal staff....The rest of that summer, the "logging the park" group therapy sessions gave way to serious planning around Paul Shepard's table....Shepard, as conservation chairman of the National Council of State Garden Clubs, had ties throughout the East....To avoid [problems] for their revered boss [they were worried about the repercussions for Gunnar Fagerlund], who was not a willing party to the logging scheme, the naturalists decided that Paul Shepard would be the front man. The plan would unfold without Fagerlund's knowledge....Shepard...not only pieced together a coalition of eastern and national groups to respond to the logging but also stepped in to confront Wirth [Conrad Wirth, director of the National Park Service] directly....Shepard had sent forty-eight long telegrams to each of the state presidents of the garden clubs. As a result, Wirth was deluged by letters and telegrams from all over the country and with similar responses from all of the conservation organizations in the country.[9]

Little more than a month after the naturalist's project began, Wirth was defending himself before conservationists and responding to a "three-page series of questions and answers about Olympic National Park's logging," written and circulated by Paul through the garden club network. The end results of this whistle-blowing episode were that all of the conservation organizations in the country were watchdogging tree-cutting policies and that top men were transferred to other places. "The Park Service then turned on Paul Shepard for disloyalty. He was officially forever banned from employment in the national parks by Regional Director Meriam himself. Shepard then became a folk hero, but that was no consolation to him since he had enjoyed being a seasonal naturalist."[10] With this finale, Paul laid to rest his seasonal naturalist work with the National Park Service; he soon found alternatives. In the summer of 1957 he taught ecology at the Audubon Center in Connecticut. The following summer with an E.I. Lilly Foundation Grant to study environmental perception by pioneers, he followed the route of the early travelers on the Oregon Trail in Kansas, Nebraska and Wyoming, much as he had done with the Hudson River School paintings, comparing ecological perceptions recorded in journal descriptions with present sites. Jane Shepard, their third child, was born to the Shepard's in 1958.

During the next three years, 1959-1962, under National Science Foundation Grants, he would be attending summer institutes, directing ecology programs at Green Oaks for high school students and high school biology teachers, and conducting research on the social structure of crows under a National Institute of Mental Health Public Service Grant. In describing the Knox years, R. Kim Chase, a colleague and friend during that time, looks back with pleasure at their shared family outings, john-boating in the Ozarks and on the Spoon River. Kim recalls that "He was a spirit, held in awe as an academician, an exciting teacher and expert in many fields. His course, then called 'Man in the Landscape,' was a must on campus. I lost track of him from 1966 to 1972, but he looked me up in California and we resumed our friendship. He never forgets his friends." Kim identifies a proclivity in Paul that remains, a deep-felt loyalty to friends, kin, and scholars who have influenced his life. John Collier, who was a guest lecturer at Knox in those years, also became a friend and mentor. Over the years Paul would travel to Taos periodically to visit Collier until his death in 1968, and to pay respects to his mentor, Sears, who died there recently, just short of his hundredth birthday.

In 1961, after a lapse of some fifteen years, Paul resumed travel and study abroad that during the rest of his life would carry him around the Planet Earth several times. He and his family spent an academic season in Wellington, New Zealand, where as a Senior Fulbright Research Fellow at Victoria University, he studied the perception of the landscape by English settlers. These findings, as well as his research on perceptions of Oregon Trail travelers, would feed into a monograph as well as a book extending his dissertation topic.[11] The end of this phase of his life took place on his way home from New Zealand, when he was notified that his father had died of a heart attack.

A recipient of an NSF Summer Institute on Advanced Biology Instruction, conducted during the summer of 1962 by the University of Southern California in Costa Rica, Paul picked up a parasite that imbedded itself in his leg. A local doctor removed it by baiting it with a lump of fat. The creature developing in his muscle was a portent of the coming upheaval within that would alter his life drastically during the next two decades. That these changes would occur shortly after his father's death is understandable. In his prime, Paul was caught up personally in the social upheaval of the sixties, and, without the watchful eye of the older Shepard, he was freed to follow the explosion of possibilities.

Paul ascribes some of the changes in his life during this tempestuous time to the writing of his book, *Man in the Landscape*.[12] In a new preface for the re-publication of the book in 1991, he wrote of the effort that went into it and the perennial revisions during the thirteen years between his thesis and its publication. He remembers the constant rewriting, "the devils who disintegrated the book's diverse topics" as well as the "inner demons who thwarted the writing of lucid English....The scope of those years [from 1950-1967]...was a part of a period" that for him "personally and for the meaning of nature in America" seem "volcanic" to him now. He goes on to say that he wishes that he could say that the book finally became all of a piece, but denies that it did. "The notion does not do justice to the price paid, that black hole of perpetual revisions in which frustration contributed to acts of personal and professional desperation."[13]

By the time the book was published, Paul would end one marriage and begin another with June Smith Atwater, a geneticist. He left a tenured position at Knox College and Green Oaks, and moved to Ashfield, Massachusetts where during the next ten years he held academic appointments at Smith College, Williams College,[14] and Dartmouth College. Struggling with the problems of a reconstituted family, he entered a chaotic, but very productive, period of his life. Although *Man in the Landscape* marked a rift in his life, it also laid out major themes that would appear in the coming years in books, dozens of essays, and guest lectures—namely the acknowledgment of our paleolithic origins, our indebtedness to animals (especially to the bear) for our human ontogeny, and the importance of primary attachment of child and nature for ecological sanity. A scholarly struggle ensued, as Shepard attempted to describe the relationship between human perception of landscape and its translation into attitudes toward nature. He speculated freely, and spoke with students as well as other scholars in seeking a deeper understanding of our environmental dilemma. Biological and historical clues merged in a complex and expanding view of the human/nature affiliation.

In 1969, working from readings he used in classes and research, Paul and Daniel McKinley, a biologist and friend from the Missouri years, published an anthology, *The Subversive Science*[15] that became a popular reader in environmental studies courses. His introduction to that book is his best known essay. The first anthology led to another, *Environ/mental*,[16] also written with McKinley.

His disinterest in administration and impatience with the traditional teaching of biology left Paul dissatisfied with his academic appointments. During the summers of 1967 and 1968, he ran courses with his war-time friend, Mal Wood, who was at the California Institute of Arts and Crafts. In 1970, Paul and June moved temporarily to San Diego, California where she received a research grant at San Diego State College. Paul, with a John Simon Guggenheim Foundation fellowship for research on the cultures of hunting/gathering people, worked on a new book and taught courses at the University of California at San Diego and Pitzer College. Although he returned to an academic appointment at Dartmouth in 1972 and 1973, by the fall of 1973, the Shepards had made their final move to California. June accepted an appointment at San Diego State College and Paul was named the Avery Professor of Natural Philosophy and Human Ecology at Pitzer College and the Claremont Graduate School, a chair fashioned for him by John Rodman, who was then dean.

Paul bought a cottage nestled in live oak in the ambience of the San Gabriel Mountains. He raised Jungle Fowl and relished the sound of their hoarse crowing echoing against the canyon walls. The comforting surround supported his research and writing. From landscape and gardens, Paul had moved to environmental thought and, from there, to hunting and gathering—where animals are the central actors in the human imagination. The year that he moved to Claremont, *The Tender Carnivore and the Sacred Game*[17] was published. He continued his travels abroad in felt kinship with nomadic hunters and gatherers, trying to document the worldwide influence of animals on human culture presented in cave paintings and ancient temples as well as modern representations. In *Thinking Animals*,[18] published in 1978, he would broach the complex entanglement of humans and animals.

~ ~ ~

In the year following his father's death, Clara moved to Denison University in

Granville, Ohio, where she became a housemother for the Kappa Alpha Theta Sorority. She enjoyed the exuberance and spirit of young college women as well as vacations with Paul and his family in Massachusetts and California. In 1979, she retired and moved to California to live with Paul. Six month later, in May, 1980 her death would usher in yet another transition in Paul's life.

At this point the relation of ontogeny and history entered as Paul's primary intellectual preoccupation. His concern for childhood development, grounded in place, care, and our evolutionary past, had trailed him throughout all of his studies. This association began during the Yale years, when Edith Cobb, upon the recommendation of Hutchinson, had written Paul and asked for his comments on her manuscript that would eventually become *The Ecology of Imagination in Childhood.*[19] Cobb's work profoundly reinforced Paul's notion of the importance of the natural world in childhood. A study of the ecology of age groups in animal populations furthered his belief that psychology had neglected ontogenesis from a biological perspective. A conference on children and the urban environment held in Washington, D.C.[20] in 1979 would also significantly influence Paul's thinking about the relationship between place and ontogeny. In this conference he participated in a symposium with Roger Hart, Margaret Mead, Harold Searles, and Yi Fu Tuan, which was possibly the major meeting on children and nature in the twentieth century. These experiences interacted with reflection on his own childhood, convictions acted out as he nurtured his children, and the birth of his first grandchild. With a Rockefeller Foundation Fellowship, he translated these diverse perspectives into an impassioned statement in *Nature and Madness,* published in 1982, which Paul believes is his most important book.

By 1984, Paul and June had parted ways. Appropriately at that time Paul returned to a book in progress that was closely tied to the idea of renewal. Throughout his tracking of animals, the bear had emerged in his work as the one whose sacredness most influenced humankind in the northern hemisphere. In 1985 he published *The Sacred Paw,* in which he uncovered the theme of the "metaphysical mother" in respect and reverence for the bear. "A feminine principle of birth, growth, death, decay and rebirth lies at the heart of the veneration of the bear, for the bear is the supreme model—and therefore the guiding spirit—of the theme of renewal."[21]

In the winter of 1985, he was on his way to India as a Distinguished Visiting Lecturer, with a stop-over at the Teton Science School in Jackson, Wyoming where he conducted a symposium on "Children and Nature." On the way, he stopped in Salt Lake City where he was a guest lecturer in a seminar I was conducting at the University of Utah. This was our first meeting. Two years before, a dream had visited me and had lured me into the Arctic:

> I was walking with a man carrying a child in my arms. Vast expanses of tundra fell away in all directions around us. The man walked slightly ahead of me, and as I followed him, I kept a wary eye on two young bears following behind. Noticing, he assured me I had nothing to fear. We

> stopped on the crest of a ridge; the sky was beginning to lighten. Anticipating a sunrise, I was surprised by shafts of light flaring up and undulating like curtains in a breeze across the entire horizon. I turned to speak to the man, but in his stead was a magnificent bear with a head so large it filled my entire field of vision. I was unafraid, assured by some primordial faith that we should be there together. We stood side by side, he on his hind legs swaying back and forth, witnessing the spectacle of the sky.[22]

Suspecting that he was the bear in my dreams, I decided to take up with Paul Shepard.

In *The Sacred Paw,* Paul began uncovering what continues to be a primary interest, the circumpolar cult of the bear in tribal and present peoples. Since the publication of his book, he has continued extensive research on the bear as well as on other animals. In two terms as a national lecturer for the Scientific Society, Sigma XI, during 1984-1988, he expanded his understandings of the place of animals in human culture.

Now as an elder, he delights in criticism of "sacred cows," as well as panaceas that erupt frequently in contemporary intellectual society. He is not at all worried about being politically correct. His interest is in understanding our affiliation and dependence on other creatures in the natural world. In a contributed chapter to Stephen R. Kellert and Edward O. Wilson's *The Biophilia Hypothesis,*[23] he questions certain kinds of proximity to animals. With a strong critique of domestication and the subsequent loss of wildness in our own lives, he states his case in no uncertain terms:

> My concern here is not the destiny of these lumpish, hand-licker-biters among humans who are desperate for the sight of nonhuman creatures because they touch some deep archetypal need....My focus is the effect of the replacement of domestic for wild animals in our psychological development, especially in the formative processes by which we mature....When animals as domestics came literally into our household...[t]here was the end of respect for the other on its own terms.[24]

Paul's forthcoming book, *The Others: Animals and Human Being,* is an homage to the profound influences of animals upon our human nature. In this book, he extends what he began in *Thinking Animals.* I predict that *The Others* will dramatically change the way serious scholars of the human/nature continuum view the creatures who share our planet.

He has not put hunters and gatherers to rest. "A Post-Historic Primitivism" in Max Oelschlaeger's *The Wilderness Condition*[25] begins what is another long-term project which Paul calls "The Eclectic Primitive: Nature and History." He insists that we must face the fact that we are still Pleistocene creatures and our genomes are ill-suited for the world we have created. In another critical essay, "Virtually Hunting Reality in the Forests of Simulacra," he takes on restoration ecology, virtual reality technology, and postmodern deconstruction. Eloquently, he turns the tools of deconstruction on modern nihilism to return us to a place in nature we have never left.

> A thousand million pairs of eyes, antennas, and other sense organs are fixed on something beyond [ourselves] that sustains [our] being, in a relationship that works. To argue that because we interpose talk or pictures between us and that shared immanence, and that it therefore is meaningless contradicts the testimony of life itself....It is the same world in which we ourselves live, experiencing it as process, structures and meanings, interacting with the same events that the plants and other animals do.[26]

The Avery chair has served Paul and Pitzer college well during these years. There he has taught interdisciplinary courses in a supportive environment. He has led students in seminars on such topics as "Catastrophe." "The Desert as a Place," a field ecology course (offered through the years with Carl Hertel), has combined ecology and art and taken classes to the deserts of California and Baja. Two of his courses, "Encounters and Confrontations with Nature" and "Animals and the Imagination," have been perennial favorites with students. He relishes a friendship with John Cobb, and the two of them have offered seminars in "Theology and Nature" at the Claremont School of Theology. Now approaching retirement, Paul continues to be a brilliant and respected teacher.

Time and a fondness for keeping on the move have carried Paul around the world several times. He has continued his research and lectures in Greece and Italy, in India as a Fellow for the International Exchange of Scholars, and, most recently, in Scandinavia. Caught in one place for long, he begins staring out of windows and looking at world maps.

~ ~ ~

A latter-day partner attempting to trace the vagaries of her spouse's complicated career, I learned more than I had intended in the writing of this chapter, yet, in closing, I am unable to adequately answer my initial question: How, indeed, does such a life come to be? Although I am thoroughly impressed by the nurturance he received during childhood and his fine up-bringing, education, and mentoring, these are only clues to his ontogeny. Perhaps the most profound influences on Paul Shepard were his hawks. They taught him to catch the thermals and to ride them high, but more importantly, to let rising updrafts intercept down-gliding.

I began this essay with Paul's high school reunion. I end with a more recent experience that may illustrate the complexity and interwebbing feeding into the psyche of this unusual human ecologist. We were recently in London in the National Museum of Natural History for the filming of a BBC (British Broadcasting Company) interview with Paul on bears. As we came into the vast domed entrance, he described how fifty years before, as a young soldier on his first leave, he had come there. At that time, the museum was closed and the door, locked. He rang the doorbell. Soon a man opened a tiny window and asked him what he wanted. Paul asked if there were an ornithologist around as he wanted to talk to someone about falconry. He was invited in. The museum had been bombed and the dome, severely damaged, had collapsed into the entrance hall. He was escorted through the rubble to the office of a zoologist named Kennier and served tea. On his desk was a picture of booming prairie chickens taken by Charlie Schwartz, a graduate student of Bennitt's at the University of Missouri, whom Paul had met the previous summer. Forty years later Paul would commission Schwartz for bear drawings for *The Sacred Paw*. Now we were waiting for BBC to set up a place for Paul among shelves containing trays of cave bear skulls, animals out of the past that had beckoned to him in his research. We were asked to wait in a curator's office where on the wall was a poster that read "Neanderthal and Proud." Seeing it, Paul shook his head and chuckled. I could imagine what he what he might be thinking: The War Years, Bennitt and Schwartz, prairie chickens, bears, Neanderthal hunters and gatherers, were all circling back on the rising currents of time.

Notes

1. I thank Paul Shepard for generously opening his files to me, answering my many questions, and correcting errors and misconceptions in the manuscript. He has shown patience and forbearance in the face of my prodding. I gratefully acknowledge editorial comments offered on a first draft by him and Kathryn Ann Morton. *Henry Moore, The Nelson-Atkins Museum of Art: Henry Moore Sculpture Garden* (Kansas City, MO: The Nelson-Atkins Museum of Art,1989) compiled by Deborah Emont Scott and Deborah Leveton, 24.
2. Paul H. Shepard, *Nature and Madness* (San Francisco: Sierra Club Books, 1982).
3. Ibid., 9.
4. *History and Families, Wright County, MO* (Paducah, Kentucky: Turner Publishing Company, 1993) 10-14.
5. Paul Shepard, *The Others: Animals and Human Being,* forthcoming. 6. Ibid.
7. Paul H. Shepard, ed., *The Pictorial History of the 493 Armored Field Artillery Battalion,* 1943-1946 (Augsburg, Germany: E. Keiser, 1945).
8. Carsten Lien, *Olympic Battleground: The Power Politics of Timber Preservation* (San Francisco: Sierra Club Books, 1991).
9. Ibid., 286-289.
10. Ibid., 289-298.
11. The New Zealand research would be published as a monograph. *English Attitudes Towards the New Zealand Landscape Before 1850* (Wellington, New Zealand: Pacific Viewpoint Monographs #4, 1969). The Oregon Trail research became a book commissioned by the Smithsonian Institute, but was

never published.

12. *Man in the Landscape* (New York: Knopf, 1967), (New York: Ballantine, 1972), (College Station: Texas A&M University Press, 1991).
13. Ibid., xxi-xxx.
14. At Williams College, he did an independent study with the son of Frank Fraser Darling, whose work had influenced him significantly and whom he had met previously when Darling was a guest lecturer at the University of Missouri and at Yale University.
15. Paul Shepard and Daniel McKinley, eds., *The Subversive Science: Essays Toward an Ecology of Man* (Boston: Houghton Mifflin, 1969).
16. Paul Shepard and Daniel McKinley, *Environ/mental: Essays on the Planet As A Home* (Boston: Houghton Mifflin, 1969).
17. Paul Shepard, *The Tender Carnivore and the Sacred Game* (New York: Scribner's, 1973).
18. Paul Shepard, *Thinking Animals: Animals and the Development of Human Intelligence* (New York: Viking Press, 1978).
19. Edith Cobb, *The Ecology of Imagination in Childhood* (New York: Teacher's College Press, 1977).
20. Paul Shepard, "Place and Human Development", *Proceedings, Symposium on Children and Urban Environment* (Washington, D.C.: U.S. Forest Service and The Gifford Pinchot Institute, 1977).
21. Paul Shepard and Barry Sanders, *The Sacred Paw: The Bear in Nature, Myth and Literature* (New York: Viking Press, 1985) (New York: Arcana Books, Penguin, 1992), 105.
22. Florence R. Krall, *Ecotone: Wayfaring on the Margins* (Albany: SUNY Press, 1994),149.
23. Paul Shepard, "On Animal Friends," in *The Biophilia Hypothesis*, ed. Stephen R. Kellert and Edward O. Wilson, (Washington, D.C.: Island Press/Shearwater Books, 1993)
24. Ibid., 287.
25. Paul H. Shepard. "A Post-Historic Primitivism" in *The Wilderness Condition: Essays on Environment and Civilization*, ed. Max Oelschlaeger (San Francisco: Sierra Club Books, 1992).
26. Paul Shepard, "Virtually Hunting Reality in the Forests of Simulacra," a paper presented at a conference on The Reinvention of Nature, sponsored by the University of California, Santa Cruz, March 1993, and organized by Gary Lease and Michael Soule.

Paul Shepard At Old Siwash

William A. Reiners

Introduction

DEEP IN MY MEMORABILIA LIES a faded color photo of two college boys, arms and legs churning, voices whooping, as they leap au naturel into the murky waters of a lake. The bank from which they have leapt is soft, tan, eroding. An oak woodland is visible in the background. The year is 1956. The boys are Tom Murphy and me, seizing the moment for a cooling plunge on a hot day at Green Oaks, a college field station owned and operated by Knox College and managed by Paul Shepard, Assistant Professor of Biology.

This photo embodies much of the time and spirit of my college days, days specially enriched by the life and work of one somewhat peculiar biology professor—the same Paul Shepard. Our leaps from the eroding bank also symbolize personal plunges Tom and I made into lives devoted to the natural environment—plunges that probably would not have taken place without the influence of Paul Shepard.

This essay is a personal memory of Paul Shepard in his early years as an assistant professor. It is told through dimmed memories of an only moderately perceptive former undergraduate. Thus, it may be more personalized fable than historic truth. On the other hand, circumstances perceived and remembered by that former student are an operational "truth" for him. I trust the reader will understand the personal filter imposed between what is remembered and what really was.

The Setting

In the mid-1950s we were a nation of a comfortable 150 million, caught up in self-admiration of our newly-discovered muscularity. We'd escaped the depression with capitalism and democracy intact—if highly modified; we'd won "the war" and were the only combatant nation that had emerged untouched. Our economy was booming, veterans were finishing up their postponed educations and starting their families. We were the "City on the Hill," God was in his place, his place was in America, and Ike was his proxy. We were full of contradictions then as now. We simultaneously saw ourselves closely connected to our rural roots, while possessing a unique mechanical and business cunning, and, at the same time being the heavy industry giants of the world. We believed ourselves to be famed for our hospitality at home and our ferocity in battle abroad. We were also a much more racist society, we had complete faith that technology would solve every problem, and most Americans viewed the environment as God's gift for our use having won it through manifest destiny.

Heavy moral responsibility accompanied us chosen people, though. The cold war was a very serious presence, fear of nuclear war was a daily reality, and we bore the sole responsibility and capacity for containing the communist devil as long as necessary.

At that time, Knox College was a small liberal arts college of about 750 students. It had been founded, along with the town of Galesburg, by a band of New York Presbyterians in 1837 to spread good word of the denomination at the time northwestern Illinois was being settled. Knox was coeducational from the beginning, and an abolitionist hot-spot before and during the Civil War. It had a fine literary tradition, some strength in the social sciences, and a marginal program in the sciences. Between 1909 and 1916, one of Knox's alumni, George Fitch, wrote a series of books about college life at "Old Siwash." "Old Siwash" was proudly used as the college's epithet for decades thereafter until it was suddenly eliminated in a spasm of political correctness in 1992. It seemed that "siwash" has negative connotations for some native American groups—connotations completely unknown by the generations of alumni who used the word. Thus, the college demonstrated its "sensitivity," proved its fitness as a liberal institution in the 90s, and demolished part of its charm and heritage. Knox was "Old Siwash" in those days, though, and the nickname helps to convey the spirit of the time.

Galesburg was a western Illinois town of about 30,000. It had a main business street, town square, well-attended churches, blocks of fine Victorian homes, and red-brick streets lined by arching elms (now gone). Galesburg was a railroad town, transected by both the Santa Fe and Burlington lines. Both lines provided passenger service powered by then-modern diesel locomotives. Besides the railroad, the town also had a number of prospering small businesses and prided itself on its economic non-reliance on the College.

Galesburg was imbedded in a thoroughly agricultural landscape of western Illinois. It was a more interesting landscape in the 1950s. Rotational farming was still practiced: oats as a nurse crop for a leguminous hay crop, followed by two years of corn. Such a rotation required livestock to convert the non-grain crops to cash. Livestock abetted nutrient cycling in the rich prairie soils. Also because of the livestock, the fields were still surrounded by sound fences or sometimes bounded by "hedge apple" (Maclura pomifera). This farming system provided excellent habitat and forage for native cottontails and introduced pheasants. The agricultural landscape was altered in the 1960s as farming became thoroughly industrialized with affordable chemical fertilizers and pesticides.

The mostly flat till plain was incised by post-glacial tributaries of the nearby Spoon River. These tributaries cut deep ravines and, where mature, planed broad floodplains below the till plain. This topography provided relief from agriculture in the form of oak forests on the slopes, and elms and silver maples on the terraced flood plains. Another kind of relief of a more somber nature was the strip mining area east of town. While destroying rich fields for coal was destructive and ugly, the strip mines did provide some interesting topography of parallel, low ridges and lakes where none had existed before. On these ridges, volunteer and planted woodlands struggled to survive and grow on the infertile overburden.

A Young Professor Arrives

In was in this setting that Paul Shepard took up his first academic position as Assistant Professor of Biology in 1954. He had served in the Army in the European theater during WW II, received a degree at the University of Missouri in English and Wildlife Conservation; and earned M.S. and Ph.D. degrees in the far-sighted but short-lived Conservation Program at Yale. I always envied him his experience at Yale, having studied with Olympian figures such as Paul Sears, George Evelyn Hutchinson and Edward Deevey, Jr. Yale was a respected center for ecology at the time and the Conservation Program encouraged creative and innovative approaches toward the environment and natural resources through interdisciplinary degrees like Paul's. It was one of the intellectual centers of environmental philosophy and ethics, built on traditions extending back to Henry David Thoreau, George Perkins Marsh and John Wesley Powell, and carried forward by John Muir, Teddy Roosevelt, William Vogt, and reinforced by development of conservation programs in the New Deal government in the 1930s. The decade of the 1950s was not one of notable environmental sensitivity in the United States; Paul was a product of a small, somewhat embattled coterie of modern ecologists and enlightened conservationists.

From this situation in New Haven, Paul returned to the pragmatic Midwest with an unusual mixture of strengths in biology, animal behavior, natural history, contemporary ecology, and art history. Given the conventional nature of Knox College, I always wondered how Paul was hired there. When Paul arrived, the Biology Department at Old Siwash consisted of three persons: Lee Furrow, a vertebrate anatomist; George Ward, a plant taxonomist; and Paul. Furrow had died before I arrived in 1955, evidently a serious loss of leadership. Nevertheless, I got the impression that members of this minuscule department were supportive of one another.

Green Oaks

Beyond Paul's teaching duties was the Directorship of a 760 acre tract of land east of town around the strip mine area referred to as "Green Oaks," the titular College Field Station. Emphasis is on "titular" because the college did not actually own the land. Rather, the land was promised to the college by Alvah Green who continued to manipulate the land in an arbitrary, poorly-informed manner until his death in the 1957. Even after Green's death, Paul was faced with an ambiguous and unsupportive stance by the College with respect to his concept of the area as an intrinsic part of the college's life.

Paul had a clear vision of Green Oaks as a scientifically sound example of good land use, especially of strip mine restoration; as a tool for bringing natural history "across the curriculum" as an outdoor laboratory for college courses, and as a locus for acceptable social and recreational use. With its mixture of arable fields, strip mine lands, a strip mine lake, restored pastures, ravine and flood plain forests, Green Oaks was an extraordinary resource providing delicious variety to the surrounding agricultural landscape. Paul put much of himself into this vision, but by the time he resigned from Knox in 1964, he remained as "baffled by Knox College's official stance toward Green Oaks as (he) was in 1954." One of my most valued papers is a copy of a memoir by Paul, "not for publication," about his experiences at Green Oaks. I hope that someday it can be published as an important insight of the young man. It appeared to me that by the early 1970s, Paul's dream for Green Oaks had been, in part, realized by his successor at Knox, Peter Schramm. Whether this ultimate final success was due to differences in personal style, or to a national change in attitude about the environment, I cannot judge.

Professor Shepard—Through A Student's Eyes

I arrived at Knox as a freshman in September 1955. My naivete about everything except wrestling and music was monumental. In retrospect, Knox was a poor place for me except for three redeeming aspects. First, the college has packed its freshman class with determined scholars to rectify disastrous publicity received from a recent scandal involving fraternities, alcohol and automobiles. These became my valued peers. Second, the faculty was utterly dedicated to the welfare of individual students. I am deeply indebted to many of them for their personal attention. The third redeeming aspect was the presence of the young professor, Paul Shepard.

As a declared biology major, I shortly met fellow freshman biology major, Tom Murphy who became a life-long friend. Tom was less naive about the college and the environment than I, and had heard about this intriguing new faculty member named Shepard who was quite an ecologist. Together we introduced ourselves to the Professor and he told us of his aspirations and activities at Green Oaks. As a result, a significant part of our weekends became invested in helping Professor Shepard implement his vision at Green Oaks. This involved spending scores of Saturdays with other students maintaining trails, cutting and moving logs to a site on Green Oaks for construction of a replicate settler's cabin, and exploratory hikes and camping expeditions throughout the property. Some of the loyal stalwarts of this group besides Tom Murphy and me were John Ruzecki, Chuck Dunn, Dick Fritz and Pete Meyer. Green Oaks was our escape from courses and campus life (cars had been banned on campus due to the scandal mentioned above), and an opportunity to practice natural history and woodcraft. We brought our families to Green Oaks when they visited us at college, and shared our secret places there with series of girl friends throughout our college careers. One such young woman generously helped me with a senior research project at Green Oaks and became terribly infected with poison ivy. I, of course, was immune, and being thoroughly crass, moved on to another girl.

It was not all manual labor at Green Oaks. Some of my dearest memories are of exploring Green Oaks and environs with Paul, Tom and sometimes others, and learning, for the first time, about geology, geomorphology, natural history and land management ethics. I don't remember Paul's mentioning Aldo Leopold, but consciously or not, he was experiencing and promoting some of the same values that Leopold described so effectively for the Sand Prairies of Wisconsin not too far north.[1] It seemed that Paul was also stimulated by May Watts' little book on reading the landscape.[2] Sometimes, these weekend outings would end with a chili supper and conversation with Paul's family at his home in Wataga, acts of generosity valued by us as gold.

If I had any illusions about my natural history skills, they were quickly dissipated in this company. I still do not have the natural history skills Paul had then and probably has sharpened to this day. Paul's inspiration led Tom Murphy and me to ask to take his course in ecology as sophomores. Paul warned us that it was a difficult subject and we should wait until we were upperclassmen. We prevailed upon him, however, and we studied ecology with him in a tutorial kind of setting—just the two of us with Paul, talking about ecology. I learned that he was right about the subject's difficulty. I deserved and got a B in that course, a humorous irony inasmuch as ecology has become my career discipline.

Paul taught a modern course centering on an ecosystem approach. With hindsight I consider that rather peculiar as he didn't seem to have an ecosystem mentality or approach. But the ecosystem viewpoint stuck with me and that is my specialty to this day. Among the auxiliary readings Paul required was a paper by one of his Yale mentors, G. E. Hutchinson. In that paper Hutchinson argued that we had an obligation to understand how the earth works as a system, just as we took an interest in how our automobile or radio work. That message had a lasting impact on me of biblical force, and became personally realized in my recent contribution towards establishment of international, global scale scientific programs involving global change issues.

I still considered ecology a kind of "hobby" subject and continued to prepare myself for medical school, as did so many other biology majors at Knox and similar places. After all, I thought, one could scarcely make a living as an ecologist; and I had to think practically about the future. Then, in our junior year, Tom Murphy and I took Paul's course on "Man in the Landscape," a product of his Ph.D. dissertation and forerunner to his 1967 book with that title. In a way, this was an even more important curricular experience than was his ecology course because of the enormous sweep of intellectual and intuitive power brought to bear on the relationship between humankind and the environment. This course appealed to our aesthetic and romantic natures as well as to our narrower scientific predilections. Paul planted seeds in our brains which grew in different ways for different people, and continue to grow as long as we live. It allowed for, and encouraged, a personal search for integration of our aesthetic selves with the more deliberately rational parts of our thought. It encouraged our study of art as an expression of human psychology and culture. It created in me a powerful "sense of place" that has grown ever since. One of his class exercises was for each of us to write an essay describing our dream home within its environmental setting. After writing these, we traded them between one another and wrote analyses of our fellow students' tastes and propensities. I wrote about a rustic place on the edge of the mountains and plains, even though I had never been west of the Mississippi at that time. On a larger scale, our present-day home in Laramie, Wyoming, resembles that kind of physiographic setting. As a result of that course I took a full year of art history and returned to musical activities abandoned since high school, both of which enormously enriched my senior year and the rest of my life.

Senior year brought the necessity to make a career commitment and I continued to wrestle with the prospect of a career in medicine or biochemistry. While I was taking the first semester of comparative anatomy—a pre-med requisite—I discovered that I loathed it. Suddenly, the realization came to me that I didn't want to do those things. I wanted to be an ecologist with the blend of science, intuition and esthetics that Paul represented. Paul, who was my advisor, didn't play a direct role in this decision, but he certainly was the major influence. Inasmuch as ecology was not a mature discipline in the 1950s, I presumed that this choice mandated an academic life of sack cloth and ashes. Another irony gradually unfolded as my career progressed, however. Ecology became not only a household word (albeit highly misused), but also a growth industry during the 1960s and 1970s so that I rode a crest of disciplinary growth and maturation. My life has been relatively comfortable—not the impoverished one that I had feared, but accepted. When I had taken my "plunge," I had expected shockingly cold water. In fact, it has been quite pleasant.

Paul must have been amused by my career choice, and by my desire to study ecology via range management at Utah State University. However, a snow avalanche tragically killed the man I had hoped to work with so that I was diverted from Utah State to Rutgers University. Evidently Paul was relieved that I was going to a basic science program at Rutgers because he told another faculty member that they had together "saved me from Utah State." That I have ended up at the University of Wyoming is yet another irony.

Through the four years of Knox, Paul Shepard was my academic advisor and we were in mostly close contact. He was always a mystery to me and other students, though, because of his distant and taciturn relations with us, and his seemingly opaque approach to relatively straight-forward issues. I, for one, could never adopt his sense of closeness with animals, although I think of him whenever I kill a deer and offer a prayer to the animal's spirit. Nor could I put as much faith in intuition and instinct that he seemed to have. In fact, I became more interested in plants than animals and remember being offended at his remark that "people that grow potatoes, think like potatoes." He had our utmost respect but we never felt that we knew the man well or really understood him. He may have been maladroit in handling students and parents. After introducing him to my family, my Mother proclaimed him as definitely "weird." For us, he was mysterious.

After Siwash

After graduating from Knox in 1959, I took a short stint in the Army (all able-bodied males faced a military obligation in those days) and finished a Ph.D. in ecology at Rutgers. I took my first position at the University of Minnesota in 1964, then moved to Dartmouth College in 1967. Paul left Knox in 1964 and we lost track of him. Meanwhile, I maintained friendships with Tom Murphy and other Green Oaks associates: one of our recurrent topics was of the mysterious nature of Paul Shepard and wondering what he was doing now.

Thus, it was an amazing surprise in 1971 when Paul showed up on the Dartmouth campus as a visiting professor in the Environmental Studies Program I had helped found. Everything now was different. Paul had a different wife, he had switched from "ivy league shabby" to "cool California beads and chains" and was a warm, personable colleague. It was thrilling to be with him as fellow professors in this new environment. Our families had dinners and outings together, we had wonderful discussions of the environment, and we played racquet ball together in the Dartmouth gym. I could scarcely believe my good fortune! Unfortunately, it couldn't last. Paul took a superior position at Pitzer College in 1973 and our ways parted again. We have stayed in better contact since that time, however. It has been a joy to see his career and reputation continue to grow.

In 1983, I left Dartmouth for the University of Wyoming, an outrageous career move on the surface of it, but motivated by an interest in another place—another life. I like to think that Paul influenced my curiosity and willingness to invest myself in other environments. The "sense of place" he presented to students back at Old Siwash does not necessarily require planting permanent roots, but rather the ability to explore and inquire about the essences of different environments: their histories and landscapes; the relationships between the people and nature; the imagery and customs of the place. Obviously, Laramie, Wyoming, has provided a rich contrast with Hanover, New Hampshire, and Minneapolis. My wife and I have explored other environments for extended periods of time as well: Washington, D.C.; Albuquerque, New Mexico; Nottinghamshire, England; Berlin, Germany; Costa Rica; Santa Barbara, California. At each locale we have come to appreciate the architecture, woodlands, history, customs and imagery of those environments. Relicts of those places are scattered throughout our home.

What of Tom Murphy, the other leaping lad in the photo? After graduate school in biochemistry at Yale and a subsequent military tour in Viet Nam, Tom joined the EPA, rose through the ranks, and presently is Director of EPA's large and nationally important Environmental Research Lab in Corvallis, Oregon. His "plunge" into the environmental business has led him to a very important position of leadership in highly practical aspects of our environmental dilemmas.

Meanwhile, Paul Shepard continues on as well. He seems to have a kind of immortality to me now, rather like the "Perfessor" had with "Johnny Shawnessy" in Ross Lockridge's romantic novel *Raintree County.*[3] The "Perfessor" keeps coming back over the years, sometimes in body, sometimes in spirit. In fact, Paul summers in western Wyoming now. I cannot speak to him about the role he has played in my life. But, maybe he will have an opportunity to read this memoir, and reflect on those early days at Old Siwash when young men sought to understand him and share some of his insight and knowledge.

Notes

1. Aldo Leopold, *A Sand County Almanac* (New York: Oxford University Press, 1949).
2. May T. Watts, *Reading the Landscape: An Adventure in Ecology* (New York: MacMillan, 1957).
3. Ross Lockridge, Jr., *Raintree County* (Boston: Houghton Mifflin Company, 1948).

Talking With Paul
Joseph W. Meeker

FOR A QUARTER OF A CENTURY, Paul Shepard and I have been in conversation. Sometimes there are pauses of several months at a time, but there is nothing especially awkward about these quiet times; when we resume, the rhythm of our talk is restored. We talk about ideas, about projects we are working on, about books we are reading, about our families, about plants and animals and places, rocks and stars and oceans. Sometimes we disagree, but we never really argue, perhaps because we both recognize that our areas of disagreement are likely to be topics where many interpretations can be reasonably held. We talk like friends, like colleagues, and occasionally like therapists or bartenders for one another.

Sometimes our conversations have been recorded, as they were several times when I was producing "Minding the Earth," a radio series carried on National Public Radio. I want to use some of those on-the-record talks as well as some of our more private conversations to convey some sense of this man and his work, and how the two are blended.

One such chat took place at Paul's home in Palmer Canyon near Claremont, California. All during the conversation Paul's jungle fowl were crowing in the background.

Joe: It's been a crazy time for the past half-century. Relationships between humans and natural systems have become strained beyond bearing. What do you think is happening?

Paul: You speak of major changes in society, and when I begin to think about these I come back to thinking about major changes in me. The only way I can answer is through reflecting on my own sense of descending into the maelstrom step by step over a period of about thirty years. I started out as a young undergraduate at the end of the second world war with an interest in birds. I thought birds were going to be my life work. I went to a university where I discovered that if you were deeply interested in birds, you should also take some interest in what birds eat, where they live, the water and soil they needed, and in what their human surroundings were like, what the people were doing to the birds. So, I got myself into the study of natural resources and the management of habitats in general. My birding suffered some. I had to go out a few less mornings, and had less contact with the birds I was interested in, but I felt the sacrifice was worth it.

Then I went off to graduate school, and I had to take another step into the descent, because there I was told that these techniques were all very well, but that policy decisions were more important. Those were being made at a level above that of the ordinary professional who was working with soil and water. We had to know something about the history of society, and how legal systems work, and something about anthropology. So, I submitted my education to a broadening effect, even though that was not the kind of thing that was normally done for advanced academic work. As a result of that, I turned my dissertation finally into a book called *Man in the Landscape*, which was an attempt to look at the history of attitudes toward nature in the last couple of centuries of European and American history.

Joe: By that time, you'd been led beyond politics toward philosophy, hadn't you?

Paul: That's where it was leading me, and of course, I was seeing fewer and fewer birds. What was happening in a somewhat larger field was that we had become sensitized as a society to the notion that not only did we have to take political and economic and social account of influences on the natural world, but now we had to begin to think in terms of ethics and morality. We had to recognize that there were religious roots involved, and that there was a deeper and still broader background in the cultural history that went back many hundreds of years. For this thinking, of course, we owe a great deal to people like Aldo Leopold, whose notion of a land ethic or an ecological conscience was important. So, I abandoned my studies of the history of esthetics and attitudes toward the landscape, and began to examine anthropological backgrounds. I thought, well, if we could be so fickle about our valuing of nature as to, in one era admire mountains, and in another era, think of them as something horrible, then obviously we were not going to solve our problems in any permanent way on that basis.

So I set myself the task of reading anthropology, seeking the advice of anthropologists and looking into the deep human past in the hope of finding there some kind of model or guide to the way in which we should behave in nature that didn't depend upon what legislatures did, or what the fashions of the times dictated about what was beautiful in nature, or what we wanted to preserve or not preserve. As a result of that, my book on hunting, *The Tender Carnivore and the Sacred Game*, emerged (and vanished quickly enough).

Joe: Not without having substantial impact.

Paul: That led me toward a strong sense that information itself was not going to make much difference. By 1970 we had not simply come alive at that moment, as it seemed to many young people, to the awareness of human relationships to nature. At least since the turn of the century, we had been very much aware that what we did in the natural world was going to determine the well-being of generations ahead of us. And yet, it's like the farmer in Missouri who said "I'm not farming now half as well as I know how." It wasn't a matter simply of getting information, or finding out about ourselves, or acquiring the data we need to know about who we are. It was not a matter, as I still somewhat naively supposed, of communicating ideas and information, but something else seemed to be at work. My own more recent sense of departure into a dark blackness in search of a kind of direction moves me away from the notion that education in its ordinary sense, producing an integrated understanding of complex systems through the sciences, social sciences, and humanities would itself bring us to a place where we were prepared to make those adjustments permitting us to live at peace on our planet.

Joe: Those "dark places" you're talking about are nooks and crannies and hidden spots in the human mind, aren't they?

Paul: They are. If we take the attitude that what we're doing on the planet is suicidal, self-destructive, in the long run; if, for reasons we don't understand, but intuit, the destruction of other species somehow will lead to the end of humanity; and if our course does not seem to be altered by learning more about these ecological systems, then there must be other reasons beyond ordinary access to information. There must be motivations that lie hidden from ourselves. We must be moving along paths that perhaps we got onto some time ago, that require a different kind of exploration of the human psyche.

Joe: It's always interesting, and a little dangerous, to ask an author about the relationship between his own personal history and the culture to which he is responsible. Do you see a parallel between the development of Western consciousness in our time and your own personal history?

Paul: Yes, I think that's exactly it. What has happened to me is not that I have been making discoveries. This is good for one's own sense of humility. I think, Joe, what you have suggested is precisely true: what happens to one is that he is shaped by his time. Perhaps the most that an ordinary scholar like myself can hope is that he will be used as a kind of organ in which the society can think about itself. Or another sense in which life as a total organism is using people in some way to adjust itself. So I would say yes, it's a very good feeling that somehow I am being a part of a use to which human consciousness is putting itself. In that sense, I feel very much a part of the times. That's why the 1970s didn't really surprise me very much. The rise of ecological consciousness that seemed to take place with Earth Day of 1970 somehow seemed normal to me. It was the logical kind of thing to happen. And what we are involved with has to do with the psychological circumstances of our species or of our culture. It is not so much a discovery by persons as it is a growing consciousness on the part of society as a whole.

However much Paul may think of himself as a part of society as a whole, it is clear to me that his mind has been far ahead of the collective consciousness. While society is now just barely able to admit that human behavior is older than the six thousand years of civilization, Paul has been exploring for decades the full implications of an ancient human past that stretches back hundreds of millennia. If our future depends to some extent upon our ability to understand the past, then Paul Shepard is one of the few pioneers showing us the origins of our minds and souls. These do not begin with agriculture, cities, or religions, but with the hunting cultures.

~ ~ ~

Paul: I remain fairly strongly convinced by what anthropology is showing us about people living under hunting and gathering circumstances. Our older vision of these people as unthinking, unfeeling savages who bashed and grunted their way through their lives, which were mainly hazardous escapes from dangerous creatures, was a fantasy and a fabrication. As societies that anthropology is still studying, and as models of some inner creatures that we all are, these people still offer us guides as to what a rich and full human life might be like. My own thinking about the degree to which we have departed, the way we have lost, is in terms of the history of the past ten or twelve thousand years. If, in the hunting and gathering peoples that have been described to us from some twenty-odd tribes we see the normal human life span being lived through a certain kind of at-homeness, a sense of being, of full identity, in a world which is good, a world in which one's personal life and society tend to affirm one another, then one of the ways to consider the distance from this that we now experience in a world in which we seem to be increasingly alien and isolated from one another and from things natural, would be in terms of episodes in our historical past that may throw some light on the normal developmental processes that bring the individual into a full maturity.

Joe: What's so good about the hunting life? What does it have that our lives lack today? What have we lost?

Paul: This is the point, of course, where the real anthropologist would be very hesitant, reluctant to make any kind of valuation that seems to compare one society against another one. None the less, a good many of these studies of people still living as hunting and gathering tribes show them to have a profound respect for the nonhuman in their lives, a sense of well-being, a lack of the kinds of anxieties and fears that seem to trouble our own time, a rich feeling of continuity with the Earth and with the place in which they live, the limitation of desires for the accumulation of things, the sense of strong bonds of kinship which are ritualized and signified in various ways, especially in which the use of plant and animal life as food and gifts are concerned. Part of our own destructiveness about the natural world includes our misapprehensions about not only animals and plants, but about so-called "savages." Even the anthropologists have not faced up to the revolution in thinking about our own past as "savages," and our own fellow human beings who are still living as hunters and gatherers. I think that there's a tremendous revolution that's going to dislodge assumptions about the nature of being human that is going to come out of that work.

Joe: It does seem that that hunting and gathering life is more deeply rooted in the history of our species. It must have a history of hundreds of thousands of years, while post- domestication life extends only ten or twelve thousand years.

Paul: The earliest evidence of domestication of plants and animals is from about ten or twelve thousand years ago, as opposed to evidence of the existence of our species, Homo sapiens, for as much as three hundred thousand years. So there have been relatively few generations passed in which the whole of written and recorded history has taken place. Which means that the kind of creatures we are now was adapted and evolved in a very different kind of life than we now find ourselves living. We are in a sense then, each of us, misfits in a world in which our own biology had not, so to speak, foreseen.

Joe: How can we get a feel for that way of life?

Paul: Think of a world in which the child goes through a long series of integrations in a tribal and clan life which culminate, sometime between his eleventh and sixteenth year, in a strenuous initiation into society that ultimately makes him feel deeply at home and deeply connected to the land he lives in. Let us suppose that this kind of experience, part of which is initiation, part of which are trials in the wilderness and quests for vision, includes a growth in his own feeling for the cosmos as a home, of the whole realm of wild and natural life as his own kin, as belonging to a complex and beautiful Earth, an infinite order in which he has a role and a place. This process of growth involves, for instance, coming to terms with an enormous diversity of wild things around him. Every day is an experience in wild things, in things whose purposes are their own, in things whose reason for being on the Earth, however linked to him and his society, is in part mysterious. Not things given as tools, or as ends in themselves, or something necessarily to be used, but things which may be part of a language about the meaning of the cosmos.

Now, imagine for a minute the difference between a life experienced as a young person in that kind of setting, and, let us say, the village of early farmers. Now the wild has been pushed back. If we think of this taking place in the semiarid Near East, the wild land had been aggressively decimated and pushed back several miles from the village. The child is now growing up in a world where he is not only surrounded by an increased number of human beings, but of a very limited number of animals and plants, each of which has been so altered by human manipulation that it no longer has the appearance and spirit of wildness.

Joe: It doesn't seem "other" any more.

Paul: He has lost the opportunity to confront the "otherness" as a child and therefore to prepare a way of confronting otherness which he has to do in religious terms as he becomes an adult. We're speaking of a normal developmental kind of thing which we all know from our own experience is one that tends to lead us, at about the time we're twelve or thirteen years old, to think about the larger questions and to seek answers to those deeper questions of the purpose of life and of the origin of life which are going to be the meat and material of our own philosophy.

Joe: Are you saying that those questions that come up with adolescence are rooted in our evolutionary and biological history, as well as in our culture?

Paul: What the culture does is to provide its own peculiar and native answers to those questions, but the normal growth of curiosity about them is a species thing, something that all human beings share now on Earth, and probably have experienced at about that time in their lives for several hundred thousand years.

Joe: But then suddenly, with the invention of agriculture and the domestication of animals, that process gets modified. And part of that long expectation of what happens when a person grows up simply doesn't happen any more.

Paul: One way of putting that would be that as a child, growing increasingly aware of the diversity of things, and with a natural interest in the names of creatures and what they do and their relationship to one another, the child is building a kind of model, a matrix, a structure. He or she is preparing a kind of system, a way of thinking about systems, that later on will provide a poetic language for talking about the cosmos as a whole. For example, the use of animals in myth as a way of talking about how things came to be, in the beginning, depends initially upon familiarization as a child with a variety of creatures and their interrelatedness. The child is building a huge body of facts, and of course the nine year old is a very factual person. He or she is interested in concrete things, in what things actually do, what they can see, what they are. There's very little poetry in the life of an eight year old. But when he then, as a twelve year old becomes newly sensitive to the way in which language can mean more than one thing, he discovers that he has a whole body of entities in language that can be used to talk about reality at another level, at the level of the religious enterprise. Virtually all the ultimate things are talked about in words that have concrete metaphorical meanings that go back to childhood. In the village life, in which the child came to be increasingly impoverished of the richness of a natural world that had its own otherness about it, it became increasingly difficult for him to shift to use that world in order to conceive of, to talk about, and to come to live in a cosmos that was beautifully rich in itself.

Joe: We've reserved a lot of admiration for that first domesticator of plants, that first tamer of animals, like the inventor of the wheel. We have been proud of those movements from a rudimentary stage of life into what we call a "civilized" way of life. What you're saying suggests that maybe we should feel a little embarrassed about those people because they removed us from a track that we need for our own mental and cultural health.

Paul: Imagine those villages of early planters and herders as they become increasingly large, into towns. They become increasingly specialized in the human occupations towns require, and increasingly organized into bureaucracies, eventually involving a priesthood, a military class, and so on. The subordination of the individual in such a society depends increasingly upon his willingness to submit to authority. He must be ready to work, slave, drudge in the fields. What I see in this is a direction that does not lead toward the sense of fully developed individuality, the sense of confidence and independence, or of a place in a cosmic order involving a great many other kinds of beings and their own ways. In these early village situations, what was desired, perhaps not consciously so, was a less mature individual, one who was ready, like a juvenile, to follow heroic leaders, and who was still influenced by what the Freudians have told us is an Oedipal fear of authoritarian figures. My suggestion is that circumstances in those times, and those village ways, did not seem to require that one come to terms with a vast, wild estate as a member of a larger system but that one come to terms increasingly with a human society which, in order to survive, required that the individual behave more and more like a child.

Joe: Then, incrementally over a long period of time, with minor losses of freedoms and powers, we began to reach a point where we disrupted some of the basic processes that we need as a species in order to fulfill ourselves.

Paul: We lost the normal developmental program of the individual, the ontogeny of the individual to move toward a confident adulthood or maturity in which a full sense of identity depends not only upon one's sense of difference from others, but on one's sense of connectedness and relatedness to them in distinct ways.

Joe: I think I see a picture of that village life and some of the things that were being gradually lost. There are later stages that look very different. In some of the more advanced stages, civilization seems to work pretty well. Religions are established, institutions are working well enough. Aren't people often working in a fairly healthy way in some of those ancient, but advanced societies?

Paul: Yes, I think that over a period of several thousand years the great theocratic, temple cities of the Near East developed. Ur, for instance, the great Sumerian city, and the other Mesopotamian centers which developed the important theology of the feminine, the model of the great Mother Earth. They developed a strong body of ritual and ceremony which integrated the individual with the Earth and the soil. If this meant subordination to a strongly authoritarian and militaristic system of government, the sacrifice was necessary. Perhaps the later stages of maturing as an adult were given up in order that a large number of people living in great city-states could get along together and could sustain the productiveness of the Earth around them. Those great temple cities existed in relatively stable circumstances for a long period of time.

The next episode in this drama involves the awakening of what you might truly call Western consciousness. The Greeks and the Hebrew prophets were extremely skeptical of these temple cities and their reliance upon the spirits of the Earth and the spirits of place. The reaction came from the ancestors of our own modern demythologized thinking. They repudiated those religions which were dependent upon the spirits of the soil and of the Earth, with seasonal rituals of an integrated sense of a great wheel turning, on which mankind played simply a part in the cosmos, with the notion that there were creative spirits residing within the Earth, and that there was a great abundance of different gods and spirits. They created a shift toward the notion of a distant, arbitrary, invisible, unseen, vengeful god whose purposes were not known to mankind, and for whom loyalty depended upon condemnation of those great Earth rituals that had been developed by these agricultural city centers. What I am describing is the repudiation on the part of the more ascetic Hebrew fathers and Hebrew prophets of those processes of mythologizing and ritualizing in which humans integrated their lives with the Earth.

~ ~ ~

Paul and I have known ourselves to be heirs of a culture that has lost its mythic, tangible connections with natural processes. We were both ranger-naturalists in the National Park Service during the 1950s, and we saw how government agencies, created to protect natural settings, were in fact compromising and damaging them. We have travelled together, trying to restore ourselves through wonderful places, camping on Wyoming's Green River, birding in the Huachuca Mountains of Arizona, fishing in the Oregon Cascades and the Wrangells of Alaska, and sailing up and down the Pacific Coast in Paul's boat. Conversations have passed between us over scores of campfires and dinner tables, usually around the theme of how the human mind fits within the natural world, or fails to fit. We have felt the pain of our own estrangement from nature and from other human beings. Between us, we have been married five times, and we have talked long and hard to help one another through each of those transitions. We know how it feels to be a little crazy, and to connect those feelings with the human past and with the neurosis that surrounds us.

~ ~ ~

Paul: Joe, you've used the word neurosis. I didn't. Partly it's because I would like to avoid psycho-medical terminology. But it's there. Let me try to clarify what I see as the evolutionary direction of our species, and the ways in which it is related to human cultures. We know that, more than any other creature, we have a long developmental growth period, technically called neoteny. We remain childlike for as much as a third of the normal human lifespan. This is an extraordinary thing in nature. It's true to some extent of our primate cousins. It's true to some extent of all larger long-lived mammals. But it seems to be, among our kind, a high specialization. Incidentally, that contradicts the notion that we are the generalized animal. I think we are highly specialized creatures, and that this long, slow developmental ontogeny that we each go through is an evolutionary adaptation. It adapts us to cultural life. It makes it possible for the individual to be enculturated, to participate in a long sequence of enfoldments and enactments or activities in which he becomes increasingly identified with his particular people in his particular place on Earth. He learns of the great chains of connectedness that go out to all the beings in his environment.

Joe: So our culture is necessarily a part of our biology?

Paul: One might say that the biological adaptation that makes culture possible includes the use of language, but that in turn is tightly scheduled into the developmental program which is part of the gene blueprint of each individual. We each, therefore, go through an extremely complicated step-by-step process of development in which there is a readiness for input, from the culture, of different things at different times. I have emphasized, in talking about the earlier agricultural society, its loss of comprehensive otherness, non-humanness. And the ideal which the Hebrew fathers gave ultimately to Judaism and to Christianity, the idea of a distant god and the repudiation of those myths and rituals associated with the Earth; these are events which, previous to them, in hunting and gathering peoples, were normally locked into the developmental end events in adolescence that brought the individual into a sense of belongingness and connectedness to the Earth on which he lives.

Joe: It may be that the environmental crisis of our time is also a mental crisis and a mental illness—that the two may be very closely interconnected.

Paul: I think it is possible to pursue this hypothesis. We can look at other epochs in the history of the West, and possibly identify events in the normal life cycle of the child and the young person which, for its own reasons, the culture of the time has found to its advantage to destroy, to obviate, to block in some way. One way to look at what it means to be civilized has to do with increasingly becoming infantile or childlike. An example from recent times is the notion of omnipotence over the whole of the natural creation. There is no lack in the last hundred years of paeans of praise for the idea that mankind will dominate the Earth. Of course that is derived from older statements within the Christian tradition, but the idea that it could in fact be done, that this Faustian impulse to control and dominate was indeed a reachable one, touches the fantasies that ordinary infants experience. It is the fantasy of absolute control over the world. Part of growing out of infancy is the discovery that one indeed is limited in power, and that one must come to terms with a world that cannot be wholly controlled. If the dream of dominating the natural world and absolutely controlling it is a dream that a society has, maybe it's to the advantage of that society to emphasize that infantile fantasy, and to do what it can, somehow empirically, in the education of the child to see that it keeps that dream of control.

Joe: It may be a short-term advantage to the society, but it's detrimental to the evolutionary development of the species.

Paul: As a naturalist, my own assumption is that the only way we reach an equilibrium with Earth as a home is to come to terms with our own limitations. You may call it humility, or just an understanding of our dependence on other creatures, and ultimately on the Earth's diversity, and on the importance of uniqueness of place, and on being surrounded by creatures other than ourselves. It means that we do not destroy, dominate, or use up the Earth, but that we limit the human use of the Earth. This was the end goal of those cultural processes to which neoteny lent itself: the slow development of the individual allowed the culture to develop something highly unique and local and to give the individual a sense of belonging which tied him in for life to a particular place and people. It opened us to vulnerabilities, to the possibility that a society, finding that more infantile behaviors (call them neurotic if you want) suited its purposes better, could devise, however unconsciously, the means of delaying the developmental processes, locking the individual into those infantile behaviors. Something like this has been going on in civilization.

~ ~ ~

Paul and I share the belief that knowledge of the past is also knowledge of the depths of the human spirit. Our talks have led us to explore evolutionary and cultural history, but we have also explored our personal pasts together. We both have ancestors who came to North America from England early in the seventeenth century, so once Paul and I found ourselves examining the naval records at Greenwich, England, in search of traces of the ships they sailed on. For both of us, wild animals are important parts of our past and present, especially bears. We have looked for bears in California, Oregon, Alaska, and even in London, where we found the bear museum that stands on the site of Shakespeare's Globe Theater. And then, of course, there were Paul's jungle fowl . . .

~ ~ ~

Joe: Paul, while we've been sitting here I've been listening to your jungle fowl. There's an odd contrast in my mind. We're in Los Angeles County, one of the extreme developments in what we call civilization, ill or well, but certainly modern, and into this setting you have introduced the ancestor to the chicken. There are animals all around us here. Is that a part of maintaining your own mental health?

Paul: Yes, I think it is. They are the birds from which the chicken was domesticated. Here in this canyon, which is still a rather wild place, the foothills of the San Gabriel mountains, there are all kinds of things that would like to eat chickens, so that keeping ordinary chickens would be quite impossible unless you kept them in cages. I enjoy having these wild, exotic, pheasant-like birds around me, and they are fully able to take care of themselves, and to be free. These birds are descendants of a group brought from Burma in 1943, and whose relatives live in the San Diego Zoo. They do quite well here. They're beautiful and interesting, and they manage to keep my environs quite lively.

Joe: Animals play a role in the development of our mental health, and I presume that's as true for you as it is for a hunter-gatherer. Why is it important to have wild animals nearby?

Paul: As a young ecologist, in the middle of the century, I was convinced, as I think many people still are, that the importance of wild things around us was that they held together those chains of energy and material flow that keep the whole biosystem, and therefore the whole environment, as a working entity, upon which we are dependent. In recent years, I've come increasingly to doubt that certain of those creatures, especially the larger and more mobile of them, are that essential to ecosystems. I realize this is somewhat heretical from an ecological point of view, but I've never seen a very good argument that the whooping crane in fact is essential to an ecosystem. Probably the same kind of thing could be said for many of the species of larger animals. If we are to articulate an argument for their continued existence, it has to be in some terms other than economic, other than the idea that they hold an ecosystem together. What the child is doing in his native curiosity about the names of plants and animals is building a means of thinking, using the animals as the objective units from which his own cognition develops. The fundamental idea of a category grows out of the name of a species. The first experience each child has with abstract thinking, is that abstraction which is the name of a kind of creature.

Joe: So taxonomy is at the root of consciousness?

Paul: Taxonomy is. That doesn't mean that it couldn't be done with other kinds of objects than creatures, but the creatures are a marvelous taxonomy because the very nature of the evolutionary system has been to clearly demark (at least in our daily experience) one kind of creature from another one and to allow us to perceive them as related groups. The notion of a hierarchic set of categories is implicit in the species system. Later on in childhood we are able to use the idea of interrelationships among the creatures as a kind of poetic or paradigmatic model of relatedness itself. In hunting and gathering peoples this means that the ecology around them—the interwoven network of different species—is a kind of example, a tangible presence around them, of the reality of an order of their own society that they carry in their heads. There is a way of thinking about orderly groupings of human beings in kinship systems which takes its cues from looking at and thinking about natural species systems.

Joe: A high degree of diversity, like that found in large, wild populations, is therefore a measure of the diversity of our own thought patterns.

Paul: That follows, because it probably was, somewhere back in the beginnings of things, the basis for the richness of our own thought patterns. D. H. Lawrence has said it much better than I can: "birds are ideas." I think he means by that that the awakening of human thinking had to do with watching animals. And it is not just in some distant past, but that same awakening takes place again in each one of us as we grow, and it requires that kind of external concrete reality, that richness of wild things around us, in order to bring to life something that is already in our own heads, and is ready for ripening when we see and watch and try to understand those natural creatures.

Joe: Is there a chance for us, given the state that we're in, immature as we are, and made so by our culture, deprived of species diversity, influenced by a long history of limiting our perceptions, to recover some of the things that we have lost? Can we reach maturity, and rediscover our own full dimensions?

Paul: I sometimes think of myself as one of those "holocaust ecologists," who look upon the decline of the Earth in some despair. As a resource conservationist, and as a human ecologist, I continue to feel that way. On the other hand, the idea that indigenous to each of us, in the very private being of each of us, is a wise person waiting to be released, is a very hopeful thought for me. It tends to turn about that whole sense of six or seven thousand years of Western dualistic alienation which has led to the mechanizing of our world. That plunges me into a black despair; there's no way of unraveling or undoing that history. But the idea that each of us potentially carries within ourselves a person who is at home in the world, and who has possibilities for feeling a strong sense of dependence and relatedness in the world, I find to be highly refreshing, and a tremendous relief. In that sense I move toward a feeling that we can make enormous changes, but that they're not going to be made through changes in ideology, or policies, or programs. They're going to be made in the ways in which we rear our children. That's the only way to unlock the person hidden inside who is the gift from our evolutionary past.

~ ~ ~

Neither Paul nor I have so far been able to release that evolutionary person hidden within us, but we continue to try together. We have a lot of talking yet to do.

A Tribute To Paul Shepard

Morris Berman

I FIRST RAN ACROSS PAUL SHEPARD'S WORK in the mid-Seventies, in San Francisco. Peter Berg handed me a copy of *The Tender Carnivore and the Sacred Game*, saying, "Have a look at this; this is a whole different way of thinking." Indeed it was. To someone caught in the traditional categories of the social and behavioral sciences, Paul's analysis of the origins of human malaise, and his vision of culture being restored to sanity, came as something of a revelation. The anthropological notion, that our humanity had to be recovered from the past if we were to have any prospect of a future, was for me a powerful idea, and made its way into the concluding pages of the book I was working on, *The Reenchantment of the World.*

Things have a funny way of turning up again, of course. A few years later I was teaching at the University of Victoria in British Columbia. *Orion Nature Quarterly* sent me a copy of a book to review: Nature and Madness, by Paul Shepard. As I eagerly read through what Paul modestly described as a "progress report," I realized that this slim volume bordered on genius. What, after all, is a genius? A genius has (at least) two characteristics: (a) He or she does not remind you of anyone else; and (b) Once they state what they have to say, the whole thing becomes obvious, even though it was previously invisible. As in the case of *Tender Carnivore, Nature and Madness* was coming out of a whole new mental world, one that was nearly unrecognizable in terms of contemporary categories of thought. For me, the key contribution was the notion of the crucial role played by a nonhuman other— i.e., the natural world—in the psychological development of human beings, particularly infants and young children. How, I thought to myself, could psychologists have missed this? For whether we are talking about Erikson or Piaget or Margaret Mahler, the formation of human identity is always seen as "naturally" taking place within a strictly human world—a world which is in actual fact narcissistic and literally nonexistent prior to the middle of the nineteenth century. Freud was indeed right—whole cultures could be neurotic—but not only for the reasons he outlined. Human ontogeny devoid of regular interaction with the animal world was a twentieth century aberration, and accounted for a lot of the craziness of the contemporary sociopolitical landscape.

In any case, I reviewed the book, later discovering that my discussion of it was one of the very few positive reviews to have appeared. As far as I could make out, most reviewers were somehow expecting "normal" social science, and could not figure out what long discussions of ecology and the natural world had to do with Object Relations Theory or Erikson's eight stages of growth. The book had suffered the proverbial fate of the study that was too far ahead of its time; but by being willing to pry up our entire civilization from a new Archimedean point, *Nature and Madness* showed how it was possible to offer the world of scholarship a dramatically new insight into the human condition.

I subsequently invited Paul to come to Victoria and give a series of lectures based on his work. This was in 1983, and the visit was an immense success. Until very recently, universities have not been particularly interdisciplinary in their outlooks; historians and anthropologists, for example, can inhabit adjoining corridors and never talk to one another. Paul's lectures drew colleagues from history, anthropology, biology, early childhood development, and perhaps a few other categories as well; and in the ensuing question-and-answer sessions, they actually debated a number of issues across disciplinary boundaries. A senior colleague of mine subsequently remarked that this was the first time he could remember this sort of thing happening, and that he hoped it would become a regular feature of university life.

Paul and I continued to correspond. By 1985, I was knee-deep in research on *Coming to Our Senses*, and when I came to my second chapter, on "The Wild and the Tame," I borrowed heavily from Paul's wonderful book, *Thinking Animals*. Following Paul, the thesis I developed was that domestication made the nonhuman other like the self, i.e. created the category of "tame," and that once that process got going, a sharp dichotomy opened up between the Me and the Not-Me. Self is then construed as good, other as evil, and the race is on to destroy or convert the other (now seen as an enemy) so as to make the world "safe." War emerges as a late paleolithic invention; and dualistic thinking, which Claude Lévi-Strauss incorrectly saw as hard-wired into the human brain, becomes the cultural norm. Paul referred to this shift as a "jarring alteration in world view," in which a more subtle and kaleidoscopic form of thought got melted down into coarse binary categories; a condition which we still live with today. What then arises, of course, is a whole category of "intermediate species"—snakes, insects, spiders and the like— which defies classification (Linnaeus dumped most of these species into one big catch-all, *Vermes*), makes us extremely uneasy, and which subsequently gets projected onto groups of people that "don't fit"—witches, heretics, Jews, blacks, the Irish, whatever—and which the dominant culture seeks to eradicate, from time to time, without ever knowing why. In one of his most vivid passages, Paul writes of these "intermediate species":

> As boundary inhabitants, they offend the necessity of clear distinction. These animals live in cracks that are the zones of separation, or under things, the surfaces between places. The flying insects swarm around our heads, near the body orifices, which are themselves marginal areas.

> Parasites on human and animal bodies seem like indeterminate forms, neither internal nor external, neither part of us nor free. Insects crawl, which is betwixt walking and swimmming....[T]hey suggest a seething, secret world of transformations, menacing in fecundity and frightening in their metamorphoses.[1]

This passage came back to me in 1992, when I was living in Germany, and went to see Peter Cohen's remarkable film on Nazi aesthetics, released in the U.S. under the title, "Architecture of Doom." Original footage was reproduced from films made by the Nazis in the late Thirties, designed to prepare the German public for the gassing of the Jews. In these films, scenes were juxtaposed of sacks of grain infested by beetles, with shots of the earlocked Jews of the Lódz ghetto. How do we get rid of these vermin, the films asked? It then showed a canister of poison gas being released in the grain warehouse. German audiences were left to draw their own conclusions.

All of this, in any case, goes a long way toward explaining the dis-ease of modern culture, the pathology, now literally anchored in the tissues of our bodies, that passes for civilization.

There are many themes that emerge in Paul's work, and formally as well as informally they have had a strong impact on a number of scholars seeking to push out into new territory. Of these themes, four stand out in my own mind as being of central importance, so let me summarize them here (the first has already been discussed above):

(1) The role of the nonhuman other, specifically the animal kingdom, in the formation of human identity.

(2) The importance of Object Relations Theory, and what is generally called pre-Oedipal development, for culture and society at large. Paul makes innovative use of writers such as Donald Winnicott, Harold Searles, Arnold Modell and others to show what healthy infancy is like, which suggests to me how such an infancy would fit into the more "diffuse" energy pattern of hunter-gatherer cultures. His hypothesis, for example, that certain aboriginal peoples may not have what is known as Transitional Objects, and that infants in these cultures very likely cathect the natural world itself (instead of just the breast), which finally shows up (e.g.) as cave painting, is a very fertile one for future anthropological research.

(3) The categorization of societies into significant types, based on how they deal with self-consciousness and self/other differentiation. This emerged most clearly in correspondence Paul and I have had over the last few years, inasmuch as I am currently writing a book on hunter-gatherers and nomadic peoples. Paul wrote that hunter-gatherers lived in a world of affinity networks (what anthropologists still occasionally refer to as "band societies"); agriculturalists seek to blunt the pain of self-consciousness by unitive trance (what Julian Jaynes called "bicamerality"); and industrial civilization offers its citizens some form of lowered consciousness, or dullardism—television, food, sex, drugs, alcohol, workaholism—to enable them to cope. While a lot of modification of this schema is obviously war-

ranted, it is not a bad first approximation to how the human race has endeavored to deal with the problems of separation and alienation.

(4) A vision of social equality that hunter-gatherer populations may have represented, and which therefore could at least theoretically be duplicated at some future point; or which may exist even today, in certain remote areas. One can call this romanticization, but when one studies James Woodburn's work on the Hadza of Tanzania, or Richard Lee's discussion of so-called levelling mechanisms among the !Kung Bushmen, one realizes that the jury is still out. As Marvin Harris has written, there was very little indication, on the eve of the neolithic era, that ten thousand years hence "our kind" (*Homo sapiens sapiens*) would be divided into paupers and billionaires. Social inequality is not natural, and accepting it as such only testifies to the degree of our cultural mystification. A few months ago, when I returned from a conference in Sao Paulo, and sent Paul a copy of an essay I had written ("The Ocean of Tears") about the dramatic extremes of wealth and poverty that confront one in Brazil, he wrote back: "The idea that those at the centers of power and wealth—however altruistic—will never permit an egalitarian reality has simply never penetrated modern thought." I took this comment as something much larger than a specific Marxist one regarding class relations; rather, it seemed to be saying that unless we begin to understand the nature of civilization itself, we shall never be able to grasp the complete unnaturalness of human hierarchical relations.

To my mind, Paul's work is itself a type of "intermediate species," in the sense described above. It is vibrant, original, and defies traditional categories. It is not the sort of thing that can easily be turned into slogans and cliches, and in that sense perhaps resembles the kaleidoscopic consciousness of an earlier age. Yet for all of these "primitive" roots (if that is what they are), the general message of Paul's work is curiously enough closest to that of the Enlightenment, which took as its motto, "Dare to Think."

Paul Shepard has done that, and the results speak for themselves.

Notes

1. Paul Shepard, *Thinking Animals: Animals and the Development of Human Intelligence* (New York: Viking Press, 1978), 100.

Nothing Pleases Me More Than To Praise

Chellis Glendinning

IN RECENT YEARS I HAVE WONDERED, after an occasional late-night dip into Lewis Mumford, if I would ever meet up with another thinker whose work competes with that of my chosen mentor. The challenge, I thought, would surely extend beyond mere intelligence, grasp of detail, or sweep of vision; the contender would have to be a writer, articulate and passionate. As reader, I would—by my own intolerance of the ecstasy of ideas—have to drop the book into my lap every few pages and breathe.

Who would guess that the earnest professor pacing back and forth before the chalkboard in Biology 101 would, twenty-five years later, emerge as Mumford's match? I confess that I don't remember much about my freshman year at Smith College. I was eighteen and the year was 1965. But I remember Paul Shepard.

He was tall, bearded, and preoccupied. Oh sure, we dissected the frogs, collected the butterflies, and debated nature vs. nurture. But there was something else. Paul Shepard was the only college teacher I ever knew to take the class into the...wilds, all the while seeding our minds with dangerously holistic notions like...ecology.

I lost touch with Shepard after Biology 101, catapulted beyond the required coursework into a lifetime focus on the social change movements that were just then erupting. Undoubtedly Shepard was preoccupied with his own movement. His first major contribution, *Man in the Landscape*, was published just two years later, and a virtual eruption of intellectual boldness followed—*The Subversive Science, The Tender Carnivore and the Sacred Game, The Sacred Paw,* and on.

Out of these, *Nature and Madness* changed my works and days. Since Smith, I had become an admirer of the great intellectual from Armenia. Mumford's brilliance carved a crystalline picture of what is wrong with mass technological civilization and our lives within it. Shepard sanctioned this view, deepened it with rare psychohistorical insight—and then went on to open the door to what could be right. I remember the moment of revelation distinctly. I was lying on the couch in my office, a luscious July breeze blowing in through the screen door, alternately reading the book and dropping it to my lap to breathe.

"In the ideology of farming," he offered, "wild things are enemies of the tame; the wild Other is not the context but the opponent of 'my' domain. Impulses, fears, and dreams—the realm of the unconscious no longer are represented by the community of wild things with which I can work out a meaningful relationship. The unconscious is driven deeper and away with the wilderness. New definitions of the self by trade and political subordination in part replace the metaphoric reciprocity between natural and cultural in the totemic life of the hunter-foragers. But the new system defines by exclusion. What had been a complementary entity embracing friendly and dangerous parts of a unified cosmos now takes on the colors of hostility and fragmentation."[1]

The unconscious is driven deeper and away with the wilderness! In one fell swoop, the psychosocial dynamic of the contemporary world revealed! I jumped up from the couch. Western civilization—with its unceasing dissections, dichotomies, and divisiveness and its endless fixes to shore up the ensuing backlash—was at root a reflection of a dissociative process, a process championed (read: rationalized) in western thought as the mind/body split. Although Shepard himself eschews the language of diagnosis, as a psychologist I could not help but view this collective split in terms of the psychological response called post-traumatic stress. The original trauma underlying this condition, I gathered from Shepard's intertwining of arrested psychosocial development with the unfolding of civilization, was our separation from our roots in the natural world; in historic terms, our domestication.

The job of the mentor is twofold. First is to seed ideas which will intermingle with a multitude of other ideas to become, eventually, original thoughts. Second is to open the student to new ways of living these ideas. Even at a distance of twenty-five years, Paul Shepard became my mentor the day I read *Nature and Madness*. I followed up on the articulation and passion of his work in my next book, laying out the relationship of the psychological experience of post-traumatic stress so prevalent among us with the socio-economic experience of domestication which defines our every experience and possibility.

Waxing into the realm of the spiritual, I confess that my new mentor also opened my inner life. From his insight, I came to see that the society I inhabit had robbed me of the most enduring experience of being alive. I decided then and there to dedicate myself to seeking recovery from Western civilization. I did this by stepping out from the encasement of the telephones and FAX machines of the world I have always known, beginning the task of "formulating—and internalizing—treaties of affiliation"[2] with the great Other, reclaiming my authentic place in Creation.

The faith of the mentored student is to carry on the work; this dedication is what distinguishes the impact of the mentor from that of the teacher. Certainly my own students are aware of the light emanating when I speak the name Paul Shepard; nothing pleases me more than to praise where praise is due. But the work is carried on in more subtle and perhaps powerful ways—just yesterday, for instance, as I was placing footprint after footprint in the muddy arroyos of these northern New Mexican desertlands, praying to the Earth with my walking.

Thank you, Paul.

Notes

1. Paul Shepard, *Nature and Madness* (San Francisco: Sierra Club Books, 1982), 35.
2. Shepard, 34.

PART II

REFLECTIONS ON SHEPARD'S IDEAS

Paul Shepard As Prophet

John B. Cobb, Jr.

PAUL SHEPARD IS ONE OF THE TRULY GREAT PROPHETS of our time. Like that of many of the great prophets of the past, his message has been too removed from the dominant intellectual and cultural world of his own time to receive a widespread hearing. But those of us who have heard, and who have, in some measure at least, understood, have had our eyes opened and been forced to see our world in a profoundly different way. And gradually that new way of looking is spreading.

I am not one who has been able simply to enter into Shepard's vision, powerful and illuminating though I have found it. To do so would have required me to abandon other perspectives and other insights that are too deeply a part of who I am to surrender. In particular it would have required me to give up my appreciation for prophecy, and thus, in a sense, for Shepard himself!

Among Christian theologians, especially, perhaps, of my generation, the prophetic tradition is of central importance. It is the tradition of criticism of what has been and is, in light of what may and should become. It is directed particularly to one's own group. Since religion in its dominant, priestly form has a strong tendency to sacralize the status quo and the powers that be, the prophetic tradition is typically anti-religious in the sense of critiquing religious practice for the sake of justice and righteousness. In its origins and in its Jewish and Christian forms the critique is in the name of God. But in the modern world its most vigorous expressions have associated "God" with that which they attacked and have appealed instead to human ideals or ideals of the human. Marxism has been the greatest embodiment of the prophetic spirit in the past two centuries.

There has been a major theological debate as to whether the power and passion of prophecy can survive indefinitely when cut off from belief in a transcendent ground of these ideals. In Marxism the dialectic of history tended to play the traditional role of God. But as time passed there was some tendency for Marxism as a movement to lose its capacities for self-criticism. On the other hand, the ideas of Marx himself continued to generate prophetic voices outside the organized movement, some of them within Christianity, where the prophetic role of Marx has been widely recognized and affirmed even when his theories were not fully accepted. This prophetic spirit is alive in Christian liberation movements in Latin American and elsewhere around the world, as well as in the civil rights and Black theology movements in the United States.

Authentic prophecy should be distinguished from social criticism in general. The latter often examines social practice in light of professed ideals to show how it is falling short. But prophecy brings to consciousness as problematic what was previously taken for granted as the way things are and even as the way things should be. In this way it critiques existing norms and brings new ideals into play.

The urbanized people of Israel supposed that their religious feasts and sacrifices were the way of pleasing their God. When Amos, in the name of God denounced their religious practices and called instead for justice to the poor, he compelled his hearers to consider a very different way of thinking of their world. He did not, of course, achieve reform in his day, but his vision has played a role in many revolutionary movements since then, including Marxism.

Similarly, the Hebrews assumed that their God favored them over other peoples and that as they came to appreciate the power of this God as the one true God, they could rest assured of their security. The prophets told them that their covenant with the one true God placed them under special obligations rather than providing them with special protection. This reversal could hardly be understood at the time, but it has provided a major, if still difficult theme, for the traditions that have incorporated prophecy in their scriptures.

Typically, prophets appeal to some actual or imagined past as a time when the evils they deplore were absent, or at least less dominant. The Hebrew prophets appealed to the pre-urban period, whether pastoral or agricultural, as one in which Israel ordered its life in a way more faithful to its covenant with Yahweh. But prophecy does not belong to a cyclical view of history that hopes for the restoration of an earlier way. It calls for a new future, to be entered by what in the New Testament is called metanoia. Translated as "repentance," this has come to mean little more than regret or remorse, but its real meaning is reversal of direction in view of a new perception of the real situation.

Perhaps the most powerful form of prophesy in this past generation has been feminism. It has problematized for most of us a great deal that we had experienced as simply the way things are. It has done so in part by arguing that society has not always been structured in patriarchal ways, so that a non-patriarchal society is a real possibility. But it has called for a movement into a new future rather than a return to prepatriarchal forms.

Since there can be no doubt that Christianity as a whole has been thoroughly patriarchal, many formerly Christian feminists have rejected Christianity. But some feminists have remained Christian. What has made this possible in many instances is the recognition that at the heart of Christianity is the prophetic tradition, that feminism is itself a part of that tradition, and that being a Christian is more a matter of commitment to that tradition than of the affirmation of any particular set of doctrines or forms of church life.

For me, an even more important movement than feminism has been the ecological one. In 1969 my eyes were opened to features of what we collectively were doing to the planet. I read Lynn White, Jr., as a paradigmatic prophet. As a Western Christian he was pointing out the terrible responsibility of Western Christianity for the continuing ecological devastation of the world, and especially for concealing this destruction from the awareness of those who thought of themselves as devoted to justice and righteousness. Indeed, he made it clear that the focus on justice and righteousness so characteristic of the prophetic tradition itself, was too narrow. He forced me to see that we theologians must repent of our anthropocentric habits of thought in the fuller sense of turning the course of our reflection around and recentering on God and the world. To do so was to recover much that had been lost in the past, but certainly not to return to any earlier form of doctrine and practice. The problems we faced were new, and they called for new ways of thinking and acting.

It was my commitment to this prophetic tradition that opened me to the message of Paul Shepard. Step by step he forced me to rethink my previous perception of history. I had taken for granted that, despite all its horrors, history involved progress. I had not focused so much on technological progress as on progress in consciousness, the development toward greater self-awareness. Eric Neumann's Jungian book on the history of consciousness had been important to me. I located the emergence of the prophetic consciousness in that context and saw Christianity as the main agency of its expansion and further development.

I was not entirely unaware of the ambiguity of this "progress." Theologians have often understood the "fall" as the historical and existential move from innocence into inner dividedness and guilt. A movement from innocence to guilt cannot be viewed as unambiguous gain. Yet it is in our guilt that the gospel addresses us, and we have thought that redemption brings us to a mode of consciousness that involves a new wholeness and is an advance over both innocence and guilt.

We have tended to read this personal experience back into history. The Garden of Eden represents the innocence of our primal ancestors. Civilization brings dividedness and guilt but also the possibility of a new wholeness or redemption. The "New Jerusalem" toward which the history of salvation moves shares some features with the Garden of Eden, but it is also enriched by the history that lies between.

The encounter with Shepard raised my consciousness as to what was wrong with this way of viewing history. By projecting the innocence of childhood on primal peoples, it ignored the fullness of the maturity that they reached in the process of growing up. It treated their wholeness as an inferior one and thus ignored its true positive value. It accepted and even affirmed the human brokenness associated with civilization as a necessary evil, necessary that is to the movement toward true human fulfillment. I saw that Christianity had been a major contributor to this depreciation of primal cultures and to the celebration of historical suffering as the necessary cost of redemption.

My initial reaction was once to again to appeal to the prophetic tradition as the basis for responding to this new critique of what I had taken for granted. But what was new for me in my encounter with the prophet, Paul Shepard, was his prophecy against prophecy. It was precisely the self- critical consciousness, which both of us associated especially with the prophetic tradition and Christianity, that worked to destroy the wholeness of hunting and gathering peoples and every approximation to that wholeness that was achieved subsequently in the course of history.

It has seemed to me that both of us have been caught in a bind. The prophetic principle had enabled me in the name of Christian faith to open myself to valid criticisms of that faith and to understand that openness, and the resultant repentance, as a move toward the redemption of history. Now I was forced to see that just this principle of critical detachment from what is was destructive of existing human wholeness. Shepard, on the other hand, while calling for a mode of being in the world that was free from this critical consciousness, could only do so by problematizing everything that most people in our own society take for granted.

In prophetic fashion Shepard brings to bear on our society a perspective that challenges all its taken-for- granted values. He does so by recalling a past in which the evils of our society were absent or much less present. He knows we cannot return to that past; so he points us toward a future which recovers some aspects of that past, but in a radically new form. This new future would bring an end to all that we have known as history.

In many ways Shepard provides a more archetypal pattern of prophesy that I had incorporated. Influenced more than I had realized by modern notions of progress, I had minimized the appeal to the past and had accented the newness of the future toward which I hoped history would move. The encounter with Shepard has forced me to rethink this. In some ways it has forced me to become more Biblical. Instead of thinking of the "fall" as a first, ambiguous step on the road to redemption, I think of it now, more purely, as the "fall." It seems today that the best that we can hope for in the future will be inferior to what we lost in that fall.

But Shepard, for his part, must acknowledge that we are, indeed, fallen. What we need today is not only to appreciate that from which we have fallen but also to find that stance most appropriate in dealing with this fallen condition. That stance cannot be one of working against the prophetic tradition. Too many people take for granted too many ideas that are leading us collectively into human and ecological devastation. These must be problematized. Visions of both a healthier past and a hopeful future are essential if there is to be the personal and social metanoia apart from which all is lost. Even if we may hope for a time when prophecy will no longer be needed, when indeed the spirit of prophecy would be destructive of what should not be destroyed, that time is not now. The prophetic tradition arose in a fallen world, and as long as the world is fallen, it will be required for redemption.

There remains for me a lingering commitment to elements in the idea of progress. I cannot simply deplore all that civilization and history have been and done. Indeed, I cannot personally wish to exist in a primal culture or a hunting and gathering society. Perhaps this simply reflects my own fallenness and false values. But there is much that history and civilization have produced out of their fallenness and in response to it that are precious to me. I realize they are part of a total movement that now threatens the very survival of humanity and of many other species. I must denounce rather than affirm the direction of that movement. But the only future for which I can truly hope will retain much that has emerged along the way.

One reason for insisting on the retaining of much that civilization has produced is that the present population of the planet cannot otherwise be sustained. Even before civilization destroyed much of the world that had provided for hunters and gatherers, the limit to the number of people who could live in that way is far below the number of people now alive. We must retain and develop much of the technology and social organization that have made the population explosion possible even while we work to stop that explosion.

The question of population size is an important one for me as a Christian theologian. Although I recognize the acute threat to other species and even to our own survival connected with the large and growing numbers of people on the planet, I cannot but believe that each human life is of great value in itself and for God. I believe that the increase of population beyond that possible for hunting and gathering societies added to the total value realized in the biosphere even as it began the process of its degradation. My goal would be the largest population that can live sustainably and enjoyably on the planet without continuing the reduction of biodiversity or further destroying cultural diversity. I derive this formulation, one that needs many more qualifications, from my understanding of God, and it complicates my picture of the hopeful future.

Perhaps my clinging to the positive values of history and civilization is connected with the fact that I cannot really imagine a future in which the need for prophecy will be superseded. That would apply only in a static situation, and I can neither imagine nor hope for that. In a dynamic world there is always the tendency of humans to absolutize forms and patterns to which they have become accustomed. As a result they resist changes that are needed. They also enter into conflict with those who have absolutized other forms and patterns. We theologians call this treatment of relative goods as absolutes "idolatry," and it is against idolatry that the prophet inveighs.

On the other hand, I can hope for a world in which the relative need for priests and prophets change. The priest takes the basic context for granted, accepts it as normative or sacred, and helps people to live through life's transitions within it. The prophet displays the falseness or destructiveness of the context. We can hope for a world in which the falseness and destructiveness of the basic ideas and practices around which society is organized are greatly reduced, a world in which acceptance of what is would not continue to have the horrible consequences that accompany it today. In such a world a Paul Shepard could be the priest he wants to be rather than the prophet he is compelled to be today.

A Man In The Landscape: Reflections On The Person As A Place

Paul Faulstich

Mexico

CONSIDER HOW THE MAN WALKS: on the desert he flows over boulders like water. If he stumbles he does so with the dignity of a bighorn picking his way across a talus slope. Paul Shepard moves across the landscape as though it is something not entirely separate from himself. The way he physically moves through the land is echoed in the way he intellectually moves through it—intricately and sensitively attuned to its topography.

As long as I have known Paul Shepard he has been one of the elders, one of the wisdom keepers from which we have much to learn. I first came to know him in the arboreal deserts of Baja. Here, he was my teacher, and he taught me much: how to kill and eat a rattlesnake as though it were a sacred being; how to honor Western Gulls by scrambling their eggs; how to savor the gusano—that beautiful little grub at the bottom of the tequila bottle. There were other, more academic lessons, but these experiential ones, these encapsulations of philosophy, are what have affected me most profoundly.

I, a neophyte naturalist, was learning how to read the country, and one of the finest philosophers in the country was my mentor. Through his quiet vision, Paul Shepard helped me to understand the importance of "place" and the need to remain connected with that which is wild. He taught me that all information is metaphorical—that what we know of the world is gained only through our constructions of it. I came to understand that there is nothing more perfect than imperfection; consider the scalloped edge of a hand hewn obsidian blade—not as precise an instrument as a stainless steel scalpel, but a deeper expression of person and place. A fine tool. A more perfect work of art.

From observing Paul move so eloquently through the desert I became transfixed with this idea that we are not only actors in a landscape, but we actually are the landscape. This essay is, in part, the outcome of what I have observed by watching Paul Shepard in Baja and elsewhere—a profound dialogue between the man and the landscape.

Alaska

On Thinking of the Mountain
(He Will Know Better Than I)

No wolves have been
To the summit—
But their cries
Are carried on great winds
And blown across the icy aretes.

In my dreaming I stood on this summit
And heard the wolves calling.
Joyous, I turned to embrace
My brother,
With a tear frozen on his cheek.

Australia

There is no geography without meaning; no landscape without person. In Australian Aboriginal cultures, "country" is an extension of "self." Among Warlpiri Aborigines—people with whom I've lived in the Central Desert—this shared identity is sagaciously articulated in the concept of Jukurrpa—the Dreaming. For Warlpiri, there is no separation between person and place, or between person and creation. People are conceived of as having been "set down" in the genesis, and all individuals carry within themselves elements of creation. Warlpiri people and Warlpiri places are laced together in a marvelous constellation of mythology.

Warlpiri have an acute sense of bioregionalism in which people, spirit-beings, natural species, and localities are viewed as interconnected. Warlpiri worldview, as bioregionally informed, transcends a dichotimization of culture and nature, and couples sentiment with place. Landscape is, for Warlpiri, an "image" of the invisible: a tangible nexus to the spiritual forces embodied within it. Not only are there practical economic bonds between Warlpiri and landscape, there are also powerful religious, social, and psychological bonds.

"Country" is a charter for living, encompassing social norms, political structures, and economic resources. Warlpiri constructions of identity and hypotheses about the self are formed with respect to this cultural reality. The extension of self onto landscape enables the articulation of personal traits in terms of graspable phenomena. Through knowledge of the land people develop into distinctive individuals, defined by their relationship with cosmologically creative forces. The self is a locus of experience, and Warlpiri perceive the self partly as a geographical entity.

Like other Aboriginal peoples, Warlpiri tell of a realm in which the earth and its creatures do not exist in their present forms. In this realm, mythological Ancestors inhabit the inchoate country and with their powers they transform the earth and create the features of the landscape. The rocks and hills, waterholes, and trees are created during this period, as are animals and tribal laws. When the Ancestors complete their wanderings they change into intangible Spirit Beings, and they dwell in special places within the land. This inescapable and power-filled ground of existence is the Dreaming. All things of the world are instituted in this Dreaming and are related in its self-creating and enduring system. Existing simultaneously in the past, present, and future, the Dreaming is a force which overlaps and envelops contemporary Warlpiri existence. It is a sacred and eternal present.

Within this realm of cosmological power, totemic Ancestors emerge from the earth and perform acts of unusual and often superhuman prowess. They also live very much like Aboriginal people live today; they hunt, dig for water, perform ceremonies, defecate, and so forth. Although these Beings are ancestral, they continue to act upon the present-day world. In this respect, Ancestral Beings and Dreamings both come before, and continue to persist in, the contemporary situation.

Ancestral Beings are both good and bad, beautiful and ugly: in addition to Rain (Ngapa) and Fire (Ward) Dreamings, for example, there are Excrement (Kuna) and Vomit (Yurlkulyu) Dreamings. Ancestral Beings not only set down the order of things (the "Law"), but they also break the Law that they establish. The same totemic Ancestor might appear as both morally commendable and morally reprehensible on different occasions of the same myth.

Human action follows this pattern as well: people usually follow the Law of the Dreaming, but they sometimes break it. When the Law is broken, the culprits are punished in accordance with Dreaming doctrine. People behave the way that they do because of the patterns established by Ancestral Beings going about their lives. In fact, Dreamings and people are genealogically linked since people share kinship categories with the Ancestors.

Because it is basic to the order of the universe, the Dreaming is the ground of Warlpiri being. Not only is landscape understood to be the material manifestation of the highest values and ideals, but it is also understood as a psychological and physiological continuance of the individual. The landscape, then, is an embodiment of both Ancestral Beings and Warlpiri persons. As a model for all that has been created, it is an integral part of Warlpiri identity.

Mythologies, some have argued, are based on "primitive" and simple-minded misconceptions of how the world works in its real mechanistic functioning. This sort of reductionistic construction, however, precludes a fertile understanding of the organicism linking nature and the human mind. Rather than a misconception of the way the world works, Aboriginal peoples have a remarkable mythconception of a patterned universe. They hold decisively to the understanding that people and creation share an identity synthesized in the Dreaming.

Warlpiri myths and rituals can best be understood in terms of places, for it is the land itself which provides the most obvious and persisting evidence of Dreaming occurrences. The physical endurance of sacred places reinforces the significance of the Dreaming. But for Warlpiri, land is more than simple evidence, it is the transfiguration of Ancestral Being. The Land is the Dreaming.

The very mention of certain place names spark emotional responses in

Warlpiri. Once, while talking with an acquaintance about his country, he avoided mentioning the name of the place where his brother had been killed. Both the brother and the place are kumunjayi; names that cannot be spoken because they relate to a deceased person. While the name of a deceased person will cause emotional trauma if spoken, my friend had no qualms about discussing his brother or the place of his brother's death in the abstract (i.e., without mentioning names). Thus, he referred to his brother either as "my brother" or "kumunjayi," and the place was mentioned only as "place longa (beyond) Karrku." For this man, the place of his brother's death is filled with familial memories as well as mythological importance. The place name itself confers significance. Contained within it are not only the activities and characteristics of the Ancestors, but also those of deceased relatives. The way in which places as well as people can become kumunjayi illustrates the shared identity of person and country. The Warlpiri landscape, then, is not only spatially and temporally external, it is ontologically significant.

The power of this landscape is felt vicariously through the people—in how men ceremoniously recite the names of sites within their countries, or in the way an elder may weep upon visiting a sacred place. People as well as places are understood as artifacts in the continuation of the Dreaming. The relationship between Warlpiri "identity" and "country" is one of part to whole—identity is inseparable from the larger framework of place. Following Warlpiri notions of patrilineal descent, adult individuals regard themselves as the offspring of Dreamings associated with their patri-clan. "This one is my daddy," a Warlpiri man explained to me as he introduced me to one of his sacred sites.

Having been initiated in the Dreaming, identity transcends the individual, but it also develops through his or her acquisition of knowledge. Warlpiri become mature individuals through experiencing the landscape and progressively acquiring knowledge of the world. Warlpiri identity matures over time until it is as palpable and enduring as the land itself.

Navigation and location are traditionally determined through mythological reference points that locate people in relation to places and Dreaming events. It is places, not people, from which direction and spatial orientation are determined. The Warlpiri way of positioning oneself in reference to places indicates the significance of mythical geography. Extrapolating from Lévi-Strauss on animal totemism,[1] places are important to Warlpiri not only because they are good to visit, but because they are good to think. "I know 'im real good," a man explained to me, "and I think 'im in my head this place."

Landscape is a condition of being; it is not something that exists in isolation from people. Individuals take on special characteristics in relation to their association with the land, for identity is rooted in country. One of the most striking experiences of my early days in Warlpiri country was having senior men approach me and solemnly tell me about their country. As an outsider attempting to establish relationships, I found this expression quite personal; the men were telling me who they really are.

In the Warlpiri language, purlka means "old man." It is a respectful term, one which confers status and acknowledges wisdom. For me it is always a privilege to watch elders locate themselves existentially in relation to the land: observing, for instance, men piercing their arm veins to draw blood for ritual adornment, deliberately facing toward their father's country as they do so. Facing toward one's paternal country is a conscious expression by Warlpiri of the continuity between Dreaming, place, and self. Warlpiri understand human blood to be a powerful element of the Dreaming. Arm-blood from initiated men is an essential component of certain rituals; it is "really strong" as Warlpiri say: imbued with spiritual strength. Spatial co-ordinates—in this case paternal country—mediate between an individual and the Dreaming. Warlpiri become mature, initiated individuals through experiencing the landscape and progressively acquiring knowledge of their world.

Warlpiri identity is embedded in the Dreaming and manifest in the land. By learning about the country through myths, and especially through the direct experiences of touching and seeing, people become, Warlpiri say, "stronger." This way of being in the world makes it possible for people to share the same spiritual existence as landscape. Through landscape, self-and-other exist as a continuous and extended entity in Warlpiri worldview. This sharing of being is emotionally and intellectually more satisfying for Warlpiri than a simple dichotomous human/environment relation. In fact, the latter is inconceivable to most Warlpiri.

But Warlpiri identity is not contained only within the physical body or the boundaries of sacred places, for it is more holographic than fragmentary. Each segment contains the whole, as is indicated by people referring to certain places in the first person. Likewise, each sacred place along a Dreaming track holds all of the power and potential of that myth. One Warlpiri man equated a Dreaming site with a pregnant woman; the site contains the Dreaming in the same way that a woman carries her child within her womb.

Through identification with Dreaming events and places, individuals are imprinted with Ancestral markings which are both tangible and ethereal. This objectification of self gives a person a set of perceptual and intellectual coordinates from which life is organized. Sanctioned in the Dreaming, a person's spiritual identity provides the context through which perceptions of the self and the world are structured. It is during ritual enactments that spiritual identity finds its most passionate expression; it is a time when individuals become the Ancestors and become the land.

For Warlpiri, identity is not only internal and subjective, but also external and objective. Natural objects are humanized, and identification with the land is expressed in the notion that humans are saturated with placeness. The landscape is a personalized reality which provides the dominant reference points for Warlpiri identity, intellect, and action. Warlpiri ontology is conceived such that an individual exists not only in relation to other individuals, but in concert with a vital and vibrant landscape, a landscape which embodies Dreaming potencies. Through knowledge of the land and the Dreaming, people develop into distinctive individuals, defined by their relationship with cosmologically creative forces.

Totemism is a medium which further enables Warlpiri to grasp values, ideals, and other qualities by objectifying them in nature. Totemic animals and people are genealogically linked since people share kinship categories with the Ancestors. The importance Warlpiri place on totemic associations became evident to me when an elder refused to shoot a wallaby—the animal of his Dreaming—that we encountered near a sacred site with which wallabies are associated. "Him bin alright," I was told, "him bin sit down here longtime. Longtime fella that one."

While Warlpiri accept the principle of physiological paternity via sexual intercourse, they also believe that spiritual animation is an essential aspect of conception. Every Warlpiri foetus is animated by the entry of Ancestral potency (kuruwarri) into the mother's womb. The place of this occurrence defines people as possessing a special relationship with a particular track of land. Totemic country, as defined through conception, is not simply external to the individual; "country" is conceived as a physical and spiritual extension of "self." These bonds of totemism are intense and permanent; people only exist in relation to places.

Warlpiri refer to individuals who share their Dreaming as "countrymen." These people do not necessarily reside in the same place, but they nonetheless share spiritual ties with a limited geographical area. As a visitor in Warlpiri country, my provisional status as "countryman" came only after months of exposure to Dreaming stories and sites. Upon returning from a sacred site, one of my classificatory fathers (a man named Jampijinpa) congratulated me: "You bin see too much (lots of) country now Jangala (my Aboriginal, or "skin," name). Ngurrju (good)." Looking to his peers, Jampijinpa continued; "Jangala's countryman now, isn't it?"

Of course, I was never really regarded as a true countryman. Countrymen share not only interests in the same area, but they share similar identities; identities initiated in the Dreaming, cemented in ritual, and objectified in the landscape.

Near the southwestern border of Warlpiri country lies Kunajarrayi, an imposing complex of sacred sites. Although Kunajarrayi proper is a specific sacred site, Warlpiri use the name more broadly to refer to the cluster of individually named and related sites in the immediate area. Kunajarrayi is the meeting place of at least eleven Dreamings.[2] It is a region pregnant with Ancestral power. The Dreaming potencies of the area, however, are symbolically encapsulated within one very sacred site. This main site is the geographical focus of periodic rituals celebrating the Ancestors associated with the entire site complex. The Kunajarrayi site consists of numerous physiographic features that are transformations of the Dreaming: a water hole is created where one of the Wolf Spider Ancestors (Mumabunda) emerges after traveling great distances underground; a long mark on a horizontal exposure of bedrock shows where a Witchetty Grub Ancestor (Ngalgari) turns itself into a Snake Ancestor (Kalangangu) after a fight and flees; large boulders are the transformed body parts of Ancestral Women that were raped and killed.

One of the Dreamings associated with Kunajarrayi concerns an incestuous man called Wadangkula. (Be careful how you use this knowledge.):

> Emerging from the ground in Pintupi country, Wadangkula—initially of the Japanangka and Japangardi subsections—travels eastward toward

> Warlpiri country. He has a tremendous sexual appetite and an enormous penis. Wadangkula's long penis gets in his way as he walks, so he "ties 'im up like belt," wrapping it around his waist, draping it over his shoulder, and coiling it around his neck. Sometimes he separates his penis from his body, and the two of them travel together, Wadangkula walking on the ground and his penis slithering below the surface. Wadangkula travels this way—sometimes with his penis tied around him, sometimes with it disarticulated—across the salt bush country, sleeping, hunting, and dancing along the way. When Wadangkula dances, his "too long prick" flops around in front of him. Upon crossing into Warlpiri country, Wadangkula changes his subsection affiliation to Jungarrayi/Japaljarri and begins to speak Warlpiri. As he nears Kunajarrayi, Wadangkula spies a woman who has just given birth. Excited by the site of her vagina, he becomes sexually aroused and sends his expanding penis to copulate with her. Being as long as it is the penis comes out the woman's mouth and she dies. Wadangkula moves on, and as he approaches Kunajarrayi he sees a group of Napangardi and Napanangka women. The women have just finished collecting witchetty grubs and are cooking them in the coals of their fire. Sensing that Wadangkula is their classificatory son-in-law and must therefore be avoided, they cover their eyes and order him to continue on. At Kunajarrayi, Wadangkula sees another party of Napangardi and Napanangka women, these ones engaged in a passionate dance. Their dancing excites him and he sends his erect penis underground to copulate with the women. The women feel the earth shaking and see it cracking as the long, hard penis moves below their feet. "Something's trouble inside here," observes one of the women, so she tells the others to go get their digging sticks. The women sharpen their sticks real well, "like crowbar," and stab the ground furiously until they feel the soft, now dead penis. The penis turns into a stone arch, but Wadangkula himself survives the attack and travels on.

The focus of the Kunajarrayi site is the large stone arch that is Wadangkula's transformed penis (ngirntl). This monolith forms a long, low rock shelter. A rock cairn beneath the arch is the scrotum. Stashed inside is a collection of sacred rocks—smooth, spherical, water-polished stones-that are the testicles (kurlurrpa) of Wallaby Ancestors. Warlpiri say that this place is "dangerous." It is filled with Dreaming potencies; it is not, they proclaim, a safe place.

Warlpiri men approach the site in single file, from the direction that the Ancestors arrive, and they sing the appropriate song. Thus, humans re-enact Ancestral travels. They carry bloodwood branches as they walk; the leaves are said to "wake up" the Dreamings, for in ritual contexts, bloodwood leaves are considered as life-giving. The Dreamings themselves exist "inside" (underground) at this site. They are, I was told, kaninjarra; underneath the surface.

Dreaming potency is experienced by way of emotional response, and virtually nothing is meaningful or beautiful outside of the concept of Dreaming. This aspect of Warlpiri reality struck me hardest when, in the company of two Warlpiri friends, I was captivated by the beauty of a setting sun. Returning from gathering ochre at a quarry by the name of Karrku, we had a commanding view of the sunset from our vantage point high on the slopes of a mesa. As the sun melted into the horizon, I watched intently as the fluid colors of sky and earth blended. I felt a sense of wholeness as I gazed out across the landscape.

My Warlpiri companions, however, showed no apparent interest in what was, to me, a remarkable sunset. Initially perplexed by what seemed to be a dismissal of something exquisite, I soon realized that when we looked at the landscape, my Warlpiri friends and I were actually seeing very different things. Whereas I was captivated by "beauty," my companions' aesthetics were informed by the mythic and personal history that stretched out before them. I knew little of this history, finding meaning instead in the striking display of land and light. In the hundreds of kilometers that lay exposed before us, my companions saw their life histories; the sacred sites to which they belong, the paths of Ancestral Beings, the good hunting grounds, the campsites, the waterholes. As though laid out for view, the land below us displayed distinct elements of mythology and Warlpiri identity. The setting sun held little interest for my companions; the continuity between Dreaming and experience proved infinitely more satisfying. For Warlpiri, the felt quality of landscape emerges from deep within the existential self.

People not only possess land, but land possesses people. Necessitating reciprocal obligations between humans and nature, this bond of mutual tenureship fosters deep sentiments toward nature. The Warlpiri notion that land both contains and is contained by people has penetrating ramifications; person and place share an identity. The country endures in Warlpiri being, and place engenders person.

Landscape, then, is a template for the modeling of Warlpiri identity. Identification with the landscape is not dissociable from the social identity of individuals, and the Warlpiri emphasis on assimilation of geographical knowledge is so integral that the landscape basically serves as a definition of the self. This is the cultured wild.

California

Paul Shepard understands that this cultured wild is an anchor of identity, solidarity, continuity, and cosmology. The natural world, he will tell you, is transfiguration of otherwise imperceptible potencies; it is the indication of evanescent Ancestors. As an enduring artifact of creation, the landscape becomes a unifying symbol within which the tension between stability and change is resolved. Landscape is symbolic and processual; it "holds" meaning, and it undergoes change precisely because it is an expression of ourselves.

The formulation of identity, like the formulation of landscape, occurs discursively. Landscapes are built into our identities such that what is perceived of the world is dependent not only upon what lies "out there," but on the actual manner of grasping experience. From the start a person is of the world and matures through their discourse with the land.

Walking with my wife and Paul Shepard through the agave forests of the Sonoran Desert we were anointed by the visit of an Ancestor. We were returning in the late afternoon from an ancient pictograph site; with each step the alluvium crunched—like sun-bleached, crumbled bones. A distant ridge sculpted the sun's light, and the huge shadow of Bear stretched across the weather worn boulders of Dos Cabesas. This Bear, here, in the low, hot desert, was as out of place as we feral Europeans. But there also was an intimate belonging: that shadow laid across the desert as a blanket of twilight; we walked across the landscape as mythological creatures. It was no accident that we were there when Bear came to visit; we are, as Paul has put it, kindred spirits.

As the sun slipped below the ridge, Bear stretched into abstraction. What remained with us was myth—earth wisdom bathed in metaphor. Out of this mythology arose the poem "On Meeting Coyote and Being Sent On My Way":

"Come Coyote,
let us live the day!"
And we spent hours
in the shade of granite boulders.
"Coyote, can you tell me
where we are?"

I will make you a map. It will show you
north. It will show you
water. It will tell you how to live.
It will say "This spot is
Sacred." It will let you stay there."

"Coyote, can I live here, with you?"

We arose with the setting sun—
the soft sky of dusk.

"You collect mesquite beans, I
will sing water into the stones."
(It is coyote who makes the turquoise blue.)

I returned in the faint light of evening.
Coyote was chanting over luminous stones.
"Coyote, these beans are for you. And I found
some wild grapes, and killed a rabbit."

"And for you, boy, I have these stones."
Coyote threw them into the night.

"See the Sky Bear? Bear is my brother.
Follow his tail to the end. This
is Polaris. Walk there."

As I left, Coyote sang to the sky;
"The night is here, and I,
I already have my supper.
Let us play . . . ."

Coyote danced until the Great Bear slipped away
and the sky over Tecate Peak
was ablaze with morning's fire.

Wanderings

Shortly after moving to Portland, Oregon, I met Lillian Pitt, a ceramic artist who traces her Native American heritage as Yakima Warm Springs. Upon our introduction she asked me a simple and wise question: "Have you walked the country?"

This was Lillian's way of learning more about me, and of welcoming me to my new home. Yes, I told her, I've walked the Columbia River Gorge; I've been to the coast; I've climbed the mesas at Warm Springs. But, I thought to myself, I've also walked the ancient coral roads of Raratonga. I have walked the Alps and the Sierra Nevada. I've walked the rainforest of Malaysia's Taman Negara. Mexico's Barancas del Cobre. Aotearoa's glacial canyons. Hawaii's Wao Kele O Puna. All of these landscapes are both entrances and exits; to leave is also to arrive, to arrive to leave. These things we know about landscapes, whether by experience or instinct. When I put this all together—the places I've walked—I get a portrait of who I am. These landscapes have informed my being. I am not altogether comfortable with this portrait; it is fragmented with a modernistic notion of travel and adventure. It lacks a certain holism and resolve. But it is me.

I have also walked the rich territory—the good hunting grounds—of Paul Shepard's intellectual domain. In this I do not despair; we rely indispensably on the elders. His is an intellect sunk in black earth, enriched with blood and divinity. It is an intellect of existential profundity. Paul will not agree with everything I have said. Honor diversity.

Notes

1. Claude Levi-Strauss, *Totemism* (New York: Harper & Row, 1964), and *The Savage Mind* (London: Weidenfeld and Nicolson, 1966).
2. The Dreamings that intersect at Kunajarrayi include Witchetty Grub, Wallaby, Snake, Nightjar, Wolf Spider, Honey Ant, Incestuous Man, Circumcised Boy, Bush-Potato, Vine, and Dentalium Shell.

Paul Shepard: Ecological Elder

George Sessions

IT IS BECOMING FASHIONABLE and (more to the point) appropriate to pay homage to our ecological elders, the wise people who have been guiding us toward ecological/spiritual futures. Books have recently been written about our American Indian and other primal elders.[1] More recent ecological elders of European extract have helped shape modern ecocentric thinking, such as Thoreau, Muir, D. H. Lawrence, Leopold, Aldous Huxley, Paul Sears, Rachel Carson, Lewis Mumford, Loren Eiseley, and Ed Abbey. And there are our contemporary wise ecological elders to whom we owe an immeasurable debt: David Brower, Arne Naess, Thomas Berry, Wendell Berry, Paul Ehrlich, Raymond Dasmann, Dolores LaChapelle, and Gary Snyder, stand out prominently.[2] Paul Shepard belongs to this group.

The Claremont Group of Ecophilosophers

Before his recent retirement as professor of human ecology at Pitzer College and the Claremont Graduate School, Paul Shepard was a member of the first recognizable "ecophilosophy group" in the United States which centered around Pitzer College, the Claremont Graduate School, and the Claremont School of Theology in Southern California. The other two key theorists of this group are the theologian John Cobb, Jr., and the political scientist, John Rodman. Their writings and activities were particularly influential during the late 1960s and 70s in laying the groundwork for the rise of ecophilosophy in North America.[3]

Lynn White's paper "The Historical Roots of Our Ecologic Crisis" (1966) linking Christian anthropocentrism and the environmental crisis tended to focus early ecophilosophical concerns on the search for a metaphysical/ethical foundation for ecological philosophy and on documenting the anthropocentric development of Western philosophy.

In May of 1970 (in the aftermath of Earth Day I) John Cobb, Jr. sponsored a major "Theology of Survival" conference at the Claremont School of Theology (reported in *Time* magazine) at which twenty Protestant scholars responded to White's thesis. Some called for a new Christian ecological asceticism: an end to conspicuous consumption, demand-creating advertising, and the establishment of a steady-state economy.[4]

Cobb produced the most philosophically interesting paper. While admitting that Christianity, even more than Judaism, resulted in a radical transcendence of human over Nature, Cobb pointed to the increasing anthropocentric subjectivity characteristic of the development of Western philosophy since Descartes. As the "reality" of the so-called "external world" was increasingly called into question by this tradition, it became impossible, he claimed, to ascribe intrinsic worth to non-human Nature. Cobb proposed a Whiteheadian metaphysics and ethics as a solution. John Cobb has become one of the most influential contemporary ecotheologians.[5]

Having been concerned since the late 1960s with tracing the anthropocentrism in the Western philosophical tradition (which led me through the Greeks, the Cartesian tradition, the American pragmatists, Santayana and Spinoza) I read the media accounts of the conference, and Cobb's analysis, with considerable interest.

Pete Gunter and the 1971 Georgia Conference on "Philosophy and Environmental Crisis"

For those theorists who have become involved in environmental ethics and ecophilosophy since the late 1970s (about the time of the inception of Environmental Ethics journal in 1979), it might be useful, in this context, to recount some of the key developments in ecophilosophy since the early 1970s.

The first ecophilosophy conference in the United States was held at the University of Georgia in February, 1971. This small-scale conference, entitled "Philosophy and Environmental Crisis" was sponsored by the philosopher, William Blackstone, Jr. At this conference, the ecologist Eugene Odum proposed the adoption of Leopold's land ethic. The papers by Blackstone and Joel Feinberg were mostly anthropocentric, proposing rights for future generations of humans and for sentient creatures. The philosopher Walter O'Briant, like Cobb, claimed that our philosophical problems stemmed from Descartes, and remarked that "it is remarkable how close Spinoza and especially Leibniz came to espousing on philosophical grounds much of what ecologists and conservationists are now urging on scientific grounds." The University of North Texas philosopher, Peter Gunter, produced the most radical (and philosophically interesting) paper. Citing Reich's *The Greening of America*, he called for a greening of philosophers. Ecology and environmentalism, he claimed, are movements toward holism. Philosophy should construct a new organic/process ecological world view that would deny infinitism, atomism, pragmatism, and mechanism.[6]

As president of the Big Thicket Association since the late 1960s, Gunter was the first American professional philosopher to combine philosophy with environmental activism. (Arne Naess combined professional philosophy with ecological activism in Norway in the late 60s). Largely as a result of his tireless efforts on behalf of protecting the Big Thicket in Texas, which continue to the present, an 84,550 acre Big Thicket National Preserve was created in 1974. In the late 70s, as a member of the citizen's advisory committee for the Texas Coastal Zone

Management Program, Gunter proposed state limits on immigration to help stem the tide of overdevelopment. In the 1990s an ecophilosophy program has grown up around Gunter, Max Oelschlaeger, and Eugene Hargrove (editor of *Environmental Ethics*) at the University of North Texas.[7]

The 1974 Claremont Conference on "The Rights of Non-human Nature"

The first major North American ecophilosophy conference was held at Pitzer College on April 18-20 in 1974. Entitled "The Rights of Non-Human Nature," it was coordinated by John Rodman, with John Cobb, Jr. and Paul Shepard on the steering committee. This three-day conference brought together many well-known ecophilosophy and environmental thinkers and writers of the late 1960s and early 70s. Garrett Hardin presented his anthropocentric "rational" managing the commons version of conservation. William Leiss represented Marcuse's neo-Marxist position on the "domination of Nature." John Lilly discussed his work with the mind of the dolphin. John Rodman discussed his concept of "ecological resistance." Paul Shepard presented his work in ecophilosophical anthropology. The Canadian naturalist John Livingston discussed the biological perversion of interspecies dominance. Roderick Nash promoted Leopold's "land ethic." Vine Deloria, Jr. developed American Indian ecological metaphysics. The psychologist Anthony Storr raised early "ecopsychology" concerns by discussing the failure of Western psychology to define the healthy human individual in terms of his/her relation to non-human nature as well as to other humans. Joseph Meeker discussed the literary "comedy of survival." Gary Snyder presented his Zen Buddhist/American Indian reactions to the issue of the "rights of non-human nature." The philosopher Charles Hartshorne, John Cobb, and the Australian biologist, Charles Birch, presented their Whiteheadian approach to ecological metaphysics. Arne Naess, whose paper on shallow and deep ecology was published the previous year, was invited but could not attend.[8]

I learned from Joe Meeker (who was teaching at UC Santa Cruz, and knew Arne Naess) about the conference. Being unpublished at the time, I sent a copy of a ponderous 60 page unpublished manuscript "The Metaphysics of Ecology" (heavily weighted toward Spinoza's metaphysics) to Paul Shepard in 1973, and asked to be included in the conference.[9] Shepard, Cobb, and Rodman decided I should respond to the Hartshorne/Birch/Cobb Whiteheadian approach. My response pitted a materialist interpretation of Spinoza against their Whiteheadian organic approach (which was wrong-headed), but I also argued (I think rightly) in favor of a biocentric egalitarian position as opposed to their hierarchical approach to intrinsic value—an "extensionist" theory of ethics with humans and their consciousness at the apex of intrinsic value. I thought this "great chain of being" approach to be ultimately anthropocentric in that it recognizes intrinsic (or inherent) value in non-humans only to the degree and extent to which they are thought to resemble humans, i.e., in possessing consciousness or sentience. This hierarchical approach also provides support for "stewardship" positions which promote the view that humans, as a result of their superiority, have a right to dominate the Earth through global management schemes.[10]

As a young, overly-enthusiastic participant, although at the same time intimidated by all the "big names" in the field, I appreciated Paul Shepard's supportive attitude towards me throughout the conference.[11]

The Aftermath of the 1974 Claremont Conference

In retrospect, I think the influence of the 1974 Claremont conference was immense in furthering the development of ecophilosophy in North America. The Canadian naturalist John Livingston, together with Neil Evernden, became an ecophilosophy group within the Faculty of Environmental Studies at York University in Ontario, Canada. Joe Meeker became director of The New Natural Philosophy program established in 1978 as part of the International College in Los Angeles. Tutors in this program included Sigmund Kvaloy and Arne Naess in Norway, John Cobb, Paul Shepard, Vine Deloria, Jr., Gary Snyder, and Dolores LaChapelle. Unfortunately, this program was short-lived.[12] I began to edit and circulate an *Ecophilosophy Newsletter*, hoping to establish and maintain contact and share ideas and publications with the growing world-wide circle of ecophilosophers. There were 6 newsletters in all: (*Ecophilosophy I* (1975) through *Ecophilosophy VI* (1984). This essentially provided a communications link among those working in the area until more professional publications appeared, such as Eugene Hargrove's *Environmental Ethics* in 1979, and Alan Drengson's Canadian ecosophy journal, *The Trumpeter*, in 1983.

Partly as a result of the stimulus of this conference, John Rodman produced a series of what were probably the most brilliant and sophisticated ecophilosophy papers written during the late 1970s. His earlier "The Dolphin Papers" (1974) had critiqued the anthropocentrism in Western philosophy.[13]

The first paper in this series of papers (written in a flurry of intellectual activity throughout 1976-77) was entitled "Four Forms of Ecological Consciousness, Part One: Resource Conservation." The first paper offered a critique of the anthropocentrism of Gifford Pinchot and the turn-of-the-century economic Resource Conservation and Development approach to Nature. The second, and perhaps most brilliant, paper ("The Liberation of Nature?") consisted of a devastating critique of ethical extensionism and the "moral/legalistic" approaches of Christopher Stone and the animal liberationist, Peter Singer. Like Whiteheadian pansychism and the "great chain of being," Rodman claimed that both Stone and Singer

> pick a quality that is conceded to be normally possessed by humans; they make it the basis for the capacity for rights; then they find it writ large beyond the human pale. Singer picks sentience and stops with animals. Stone picks consciousness as well as sentience and suggests that it may well be present in all natural "objects." Of course, there is a pecking order in this moral barnyard....The process of extending rights to nonhumans conveys a double message. On the one hand, nonhumans are elevated to

> the human level by virtue of their sentience and/or consciousness; they now have (some) rights. On the other hand, non-humans are by the same process degraded to the status of inferior human beings....[Rodman asked rhetorically] "Is this, then, the new enlightenment—to see non-human animals as imbeciles, wilderness as a human vegetable?"

As an alternative to hierarchical, extensionist theorizing based on the criteria of consciousness or sentience, Rodman claimed that "all natural entities (including humans) have intrinsic worth simply by virtue of being, and being what they are." He proposed that, unlike Stone and Singer, we really "think the unthinkable" (citing Paul Shepard in *The Tender Carnivore and the Sacred Game*, 1973) by imagining "an alternative world in which the situation is reversed, a world in which human population and economic sprawl are reduced to a point where people live again within city walls (boundaries, not frontiers), surrounded by free wilderness. . . ."[14] (In a paper describing his own preferred form of ecological consciousness, Rodman advanced John Stuart Mill as the exemplar of what he called Ecological Resistance.[15])

During the summer of 1981, as a visiting fellow at the Research School of Social Sciences at the Australian National University, Rodman presented a paper summarizing his ecophilosophical analyses throughout the 1970s (entitled "Four Forms of Ecological Consciousness Reconsidered").[16] It is fitting that Rodman presented his paper there, since it could be argued that Richard Routley (Sylvan) of Australian National University published the first professional philosophical paper, based upon Leopold's land ethic, of the "environmental ethics" genre ("Is There a Need for a New, An Environmental Ethics?," 1973).[17]

With the establishment of Hargrove's *Environmental Ethics* journal in 1979, the philosophical field of environmental ethics officially made its debut. Early papers in the journal by J. Baird Callicott and Holmes Rolston, III, laid out the ecocentric environmental ethics position, based (like Routley) upon Leopold's land ethic. Callicott soon published a critique (three years after Rodman's critique) of the animal liberation position.[18]

If professional academic philosophers in the late 1970s felt at all comfortable with philosophical theorizing about ecology or non-human realms, they tended to feel most comfortable with an "environmental ethics" and/or animal liberation/rights approach in that these approaches "fit" into established academic disciplines: either as a type of "applied ethics" or, in the case of animal liberation/rights, as an "extension" of established ethical theories such as utilitarianism or "rights" theory. For example, Peter Singer's version of animal liberation was essentially an elaboration of Bentham's 19th century utilitarianism.[19] Many philosophers who became interested in environmental ethics in the late 70s were apparently unaware of the more comprehensive metaphysical ecophilosophical approach developed throughout the 1970s. For example, one recent textbook writer has claimed:

> When I became interested in environmental ethics in the late 1970s, few

> articles and books existed in that area....Traditionally, the debate in environmental ethics in the West has been between anthropocentric ethics and sentience-based ethics. The question...was whether intrinsic value was to be confined to rationally self-conscious agents or extended to creatures possessing sentience.[20]

Animal rights/liberation theorizing and activism has dramatically raised consciousness and concern about the mistreatment of both domestic and wild animals for people in modern industrial societies, and has served, for some, as a stepping stone to more comprehensive ecological concerns. But by claiming (erroneously) to be a non-anthropocentric approach, it turned out to be essentially a red-herring in the search for an ecocentric environmental ethic or ecophilosopy.

A minor debate did break out in the early 1980s over whether a new "environmental ethic" was adequate as a basis for a resolution to the environmental crisis, or whether what was needed was a more comprehensive ecophilosophical approach involving a major change of consciousness or "paradigm shift" (a radical change of our world view, or what Arne Naess calls a "total view"). But most professional philosophers at the time seem to have missed the significance of this issue.[21]

In his 1981 Australian paper, Rodman characterized the first three forms of ecological consciousness as (1) Pinchot's Resource Conservation; (2) Muir's Wilderness Preservation (which he interpreted as primarily a religious/esthetic movement); and (3) the Moral Extensionism of John Lilly, Whiteheadian ecophilosophical theorizing (Hartshorne and Cobb), and the animal rights/liberation movement. The fourth form (which he endorsed) he called Ecological Sensibility, substituting Leopold for Mill as his exemplar of this contemporary form of consciousness.

Rodman had earlier claimed that "the progressive extension model of ethics, while holding out the promise of transcending the homocentric perspective of modern culture, subtly fulfills and legitimizes the basic project of modernity—the total conquest of nature by man." Correspondingly, he argued that "we may need to become less moralistic and less legalistic, or at least to become less fixated at the moral/legal stage of consciousness."[22] Contrary to the efforts of Routley, Callicott, Rolston, and others to treat Leopold's land ethic as an early version of a professionally developed "environmental ethic," Rodman interpreted Leopold's *Sand County Almanac* as consciously designed by Leopold to bring about an ecocentric perceptual shift, or "paradigm change"—as a guide to the development of ecocentric consciousness. Rodman claimed that "obviously, I believe that those who see Leopold's land ethic as a mere extension of conventional ethics are radically mistaken."[23]

I think that Rodman is essentially correct in his interpretation of Leopold's intentions and the status of the "land ethic," although I have problems with his historical interpretation of the "earlier forms" of ecological consciousness as a "progressive development" (or as stages leading toward) the development of contemporary Leopoldian ecological consciousness or "sensibility." For example, Rodman did not have access to the post-1980s scholarship on either Thoreau or

Muir which has firmly established their positions as being ecocentric.[24] Therefore, Muir should not be thought of primarily in terms of an "esthetics of nature" position, although the Sierra Club, as Rodman rightly points out, had strayed from Muir's original religious-ecocentric orientation to an esthetics-recreation position.[25]

As the conservation/environmental movement moved away from arguments for nature protection based upon esthetics, economics, and resource conservation to a less anthropocentric and more holistic, ecological approach, Leopold's secular scientific approach and ethics came more and more into favor. This move on the part of the conservation community was, to a certain extent, pragmatic, although many of the newer conservationists began to think essentially in these somewhat detached secular scientific terms. In a scientific/technological and economically oriented society, it was thought that only scientific/ecological and economic arguments could win the conservation battles. But something crucial was lost in this process. For all his obvious merits and ecological stature and insight, Leopold seems a bit too secular, a bit too "civilized" and "genteel" and, like many contemporary scientific ecologists, perhaps a bit too "professorial" and scientifically detached from the wild to be seen as an exemplar of "post-modern" ecological consciousness.[26] Thoreau and Muir were good scientists too, but their religious pantheism and their essential wildness (their deep identification with wild Nature, also found in D. H. Lawrence and Robinson Jeffers, and in such contemporaries as Dolores LaChapelle, Arne Naess, Paul Shepard, and Gary Snyder) seem more characteristic of "post-modern" ecological consciousness. It is significant that one of Muir's criteria for assessing people was in terms of how much "wildness" he could sense in them.

Rodman's progressive model of the development of contemporary ecological consciousness (a model also promoted by Roderick Nash) further breaks down as we have come to appreciate the highly developed religious ecological consciousness, wildness, and egalitarian social structures of many traditional American Indian and other primal cultures. Max Oelschlaeger argues that we moderns must recultivate "Paleolithic consciousness." This seems appropriate as a criterion for post-modern ecological consciousness. In making this claim, Oelschlaeger was heavily influenced by Gary Snyder, and by Paul Shepard's writings on primal cultures.[27]

Paul Shepard, Cynegetic Culture and Ecopsychology

Paul Shepard became one of my "ecological elders" beginning in the late 1960s, along with Muir, Dave Brower, Raymond Dasmann, Lynn White, Jr., Paul Ehrlich, and Gary Snyder. In the *Ecophilosophy V* newsletter (1983) I featured a section on Shepard on the occasion of the publication of his *Nature and Madness* (1982), claiming that Shepard "is one of the true mavericks and original thinkers in the ecophilosophy area.... Those doing serious research in ecophilosophy cannot afford to ignore Shepard's writings."

With his ecophilosophical writings in the late 1960s Shepard provided some of the initial impetus for the Claremont group. His first book, *Man in the Landscape* (an extensively reworked version of his Ph.D. thesis at Yale) was published in 1967.[28] Here he was concerned primarily with esthetics, the philosophical and theological history of anthropocentrism in the West, and land-use problems. For instance, he found the prevailing existentialist and humanist philosophical alienation from nature disturbing: "By valuing only the unique and individual they rightly oppose mass man and the treatment of human beings as replaceable machines—at the price of ecological nihilism. It is the act of a screaming and demented oyster" (xiv). His chapters on "Fellow Creatures" and "Varieties of Nature Hating" are perhaps the most philosophical. In the former he criticizes Albert Schweitzer's Christianized "Reverence for Life" principle as what we would now see as a forerunner of the animal liberation position. In the latter chapter, he discusses the anthropocentric Judeo-Christian and Greek roots of human/Nature dualism in the West.

Shepard was co-editor of two ecological anthologies published just before and after Earth Day 1.[29] The first anthology, *The Subversive Science* (1969) was the more successful of the two, and probably the best ecological anthology to appear during the period of the late 1960s-early 70s. I was deeply influenced and moved by Shepard's introductory article, "Ecology and Man—A Viewpoint" when it appeared. It is arguably one of the three classic ecocentric papers of the 1960s (along with Lynn White Jr.'s "The Historical Roots of Our Ecologic Crisis" and Gary Snyder's "Four Changes").

In "Ecology and Man," he begins by distinguishing between two views of self: as "an isolated thing," and as deeply immersed and interrelated with the natural environment. In this latter (more Naessian "ecological self") view, Shepard, like many other ecologists of the 1960s, was influenced by Alan Watt's Zen Buddhist views.[30] As in *Man and the Landscape*, there is again a discussion of Greek and Judeo-Christian anthropocentrism, and misuses of Darwinian theory which leads us to see Nature as evil, or as primarily "resources" and "scenery" for humans. Foreshadowing the themes in Nature and Madness (1982), Shepard claimed that

> a kind of madness arises from the prevailing nature-conquering, nature-hating and self-and world-denial. ...It can be seen in the behavior of control-obsessed engineers, corporation people selling consumption itself, academic superhumanists and media professionals fixated on political and economic crisis; neurotics working out psychic problems in the realm of power over men or nature, artistic symbol-manipulators disgusted by anything organic.[31]

And referring to the ecologist Paul Sears (Shepard's mentor at Yale) and his characterization of ecology as a "subversive subject," Shepard popularized the idea of ecology as the "subversive science" (the title of the anthology). Shepard claimed that

> the ideological status of ecology is that of a resistance movement. Its Rachel Carsons and Aldo Leopolds are subversive (as Sears recently called ecology itself). They challenge the public or private right to pollute the

> environment, to systematically destroy predatory animals, to spread chemical pesticides indiscriminately.... They oppose the uninhibited growth of human populations, ...the extinction of species of plants and animals, the domestication of all wild places...and most other purely engineering solutions to problems of and intrusions into the organic world.[32]

While not underrating the importance of the sciences, including the science of ecology, Shepard nevertheless claims (like Naess) that "there is an ecological instinct [or intuition] which probes deeper and more comprehensively than science, and which anticipates every scientific confirmation of the natural history of man." Elaborating upon this point, Shepard pointed out that

> although ecology may be treated as a science, its greater and overriding wisdom is universal. That wisdom can be approached mathematically, chemically, or it can be danced or told as a myth. It has been embodied in widely scattered economically different cultures. It is manifest, for example, among pre-Classical Greeks, in Navajo religion and social organization, in Romantic poetry of the 18th and 19th centuries, in Chinese landscape painting of the 11th century, in current Whiteheadian philosophy, in Zen Buddhism, in the world view of the cult of the Cretan Great Mother, in the ceremonials of Bushman hunters, and in the medieval Christian metaphysics of light. What is common among all of them is a deep sense of engagement with the landscape, with profound connections to surroundings and to natural processes central to all life.[33]

Shepard concluded that "affirmation of its own organic essence will be the ultimate test of the human mind."

In his quest for an answer to the question "why do men persist in destroying their habitat?," Shepard was drawn to the evolutionary roots of humanity in our Pleistocene hunting/gathering ancestors and their "profound connections to surroundings" in his next book, *The Tender Carnivore and the Sacred Game* (1973).[34] I think this is Shepard's most powerful book; although the writing style is elegant and academic, it is initially somewhat difficult to penetrate and demands close attention from the reader.

Shepard asks, "Can we face the possibility that hunters were more fully human than their descendants?" The contemporary global ecological crisis, he claims, has been 10,000 years in the making:

> As agriculture replaced hunting and gathering it was accompanied by radical changes in the way men saw and responded to their natural surroundings. Although hundreds of local forms of farming developed,...they all shared the aim of completely humanizing the earth's surface, replacing wild with domestic, and creating landscapes from habitat.[35]

Shepard holds that most forms of farming are a disease and that domestic animals are "genetic goofies." An ecologically sane future, he argues, must come to grips with the fact that humans are genetically hunters and gatherers programmed for wild environments. It is necessary for our physical and emotional well-being that we reincorporate into our lives the basic features of cynegetic society (rituals, diet, exercise, close contact with the wild). Shepard counters criticism of these ideas by claiming:

> Most people seem to agree that we cannot and do not want to go back to the past; but the reason given is often wrong: that time has moved on and what was can never be again. The truth is that we cannot go back to what we never left. Our home is the earth, our time the Pleistocene Ice Ages. The past is the formula for our being. Cynegetic man is us.[36]

Ecologically, he proposed that the centers of the continents be returned to the wild. The book contains prescriptions for properly raising our children in future cynegetic cultures, a concern which has increasingly preoccupied his attention.

> I believe that every child under ten has three ecological needs: architecturally complex play space shared with companions; a cumulative and increasingly diverse experience of non-human forms, animate and inanimate; and occasional and progressively more strenuous excursions into the wilderness where he may, in a limited way, confront the non-human....The collection and study of plants and animals is more important than any other learning activity....In the cynegetic society [the child] would be introduced gradually to the great wilderness, which...would occupy the centers of islands and continents, unmodified by human action....From the age of thirteen the adolescent youth would move into a series of increasingly extended and arduous expeditions for which childhood and juvenile skills and his knowledge of natural history had prepared him.[37]

Rene Dubos, whose Christian "humans over nature" stewardship views were very influential at the time, reviewed Shepard's book in *Time* magazine. Needless to say, Dubos, an admirer of European totally humanized and domesticated environments, found Shepard's cynegetic proposals unrealistic and impractical.

Shepard continued to explore the human need for wild plants and animals in *Thinking Animals* (1978).[38] This is a fascinating detailed discussion of the relation of animals to humans, and animal mythology, throughout history. (This topic was discussed further in Shepard's book on the bear, *The Sacred Paw* (1985), written with Barry Sanders. As a result of problems with Sanders, Shepard later claimed, in a circulated letter, that he had written the entire book except for chapter 6—on the bear in literature—the sole chapter Sanders had written.)[39]

In *Thinking Animals*, Shepard argues that there is a natural psychogenetic development for humans. Normal human ontogeny involves a stage at which there must be an intimate bonding and relating to wild animals and the wild. Wilderness and wild animals must be protected to allow the opportunity for humans to proceed through their normal childhood developmental processes. He claims that "the human mind needs [wild animals living in unmanaged wild habitats] in order to develop and work. Human intelligence is bound to the presence of animals." This argument sounds anthropocentric, but, in actuality, it points to the inseparable bonds between humans and wild nature. In a more explicitly ecocentric vein, Shepard additionally claims, like Rodman, that the very existence of wild plants and animals also justifies their protection.[40]

Over the last several years, there have been meetings in the San Francisco Bay Area in which psychologists have discussed the new field of "ecopsychology" arising largely from Theodore Roszak's latest book, The Voice of the Earth (1992). Roszak claims that:

> The sanity that binds us one to another in society is not necessarily the sanity that bonds us companionably to the creatures with whom we share the Earth. If we could assume the viewpoint of nonhuman nature, what passes for sane behavior in our social affairs might seem madness. But as the prevailing reality principle would have it, nothing could be greater madness than to believe that beast and plant, mountain and river have a "point of view." We think that sanity—like honor, decency, compassion—is exclusively a social category.

Roszak further claims (in a section called "Ecopsychology: The Principles") that "The core of the mind is the ecological unconsciousness. For ecopsychology, repression of the ecological unconsciousness is the deepest root of collusive madness in industrial society; open access to the ecological unconsciousness is the path to sanity."[41]

Similarly, Warwick Fox's recent writings have provided a major impetus for the development of ecopsychology. Fox's theoretical ecopsychology draws upon Naess' concept of the "ecological self," which involves the processes of ever widening identification of the self to include not only humans, but wild plants and animals and the entire non-human world. Unfortunately, Fox feels that we should abandon Deep Ecology as a concept in favor of his type of ecopsychology: Transpersonal Ecology.[42]

While the books by Roszak and Fox are key contributions to the development of an ecopsychology, there is little or no explicit discussion of the importance, and the actual experiential processes of, bonding with the wild. The writings of Gary Snyder and Paul Shepard, on the other hand, can be seen as a continuous process of developing a detailed "ecopsychology of the wild." Snyder's writings, and his development of a "mountain Zen Buddhist/American Indian practice" over the years have been directly involved with issues of Nature spirituality, "ecopsychology," and bioregional living with the wild. And this is true as well of Arne Naess's development of an ecosophy, and work with the metaphysics and psychology of Spinoza, and of Gandhi. Naess's most explicit work of ecopsychology appears in his 1986 paper on the "ecological self."[43] Paul Shepard's *Nature and Madness* can perhaps be viewed as the first explicit treatment of the "psychology of the wild."

By viewing modern European history from the standpoint of ecopsychology, it makes sense to interpret Rousseau and the Romantic movement, and various nature writers to the present day, as beginning the process of developing an "ecopsychology" by pointing to the psychological importance of relating to wild Nature, and thereby suggesting a psychological and spiritual deficiency in urban people who fail to do so.[44] For example, John Muir, upon arriving in California in 1867, claimed that he couldn't find a sane person in the city of San Francisco, and promptly left for the wilds of the Sierra. "Going to the mountains," he claimed, "was going home."

Some ecophilosophers have recently begun to seize upon, and explore the implications of, Thoreau's radical and unprecedented 1851 statement that "In wildness is the preservation of the world...[and] in short, all good things are wild and free." This statement is the fountainhead of modern ecophilosophy and ecopsychology in suggesting that the modernist project of domesticating and destroying the nonhuman world, together with the corollary process of humans living domesticated lives out of touch with the wild in artificial humanized environments, is resulting in irretrievable disaster. These issues also concerned Huxley and Orwell. But the naturalists and social prophets (from Thoreau and Muir, to Huxley, Orwell, Shepard and Snyder) were also well aware that the antidote to our urban domesticity does not lie in temporarily "getting away" from the cities by vacationing in remote wild environments with our urban mind-sets intact, in the form of fast-paced high-tech commercialized egoistic adventures and outings. With this approach we continue to remain "natural aliens." This turns the wilderness into "scenery," or an "outdoor gym"—an extension of urban-industrialism and the backdrop and "scene" for competitive outdoor adventure "projects" which gets in the way of spiritual and ontogenetic bonding with the wild.

Many modern urban environmentalists see the issues surrounding the protection and promotion of wildness (both human and nonhuman) as peripheral and less important than industrial and urban pollution problems. For example, at a major ecofeminist conference in Los Angeles in 1987, some urban ecofeminist intellectuals expressed their opinion that wilderness experience and living, and promoting and protecting wildness involved little more than "macho male bonding rituals" and attempts to "play Daniel Boone." These urbanites overlook their roots in the intimate relation to the wild, as this has been articulated by such writers as Mary Austin or Rachel Carson. Carson's essay "A Sense of Wonder" stresses the importance of helping children develop nonegoistic bonding with wild Nature.

Anticipating many of Shepard's themes concerning our genetically based ontogeny which involves bonding with the wild, the botanist Hugh H. Iltus made a powerful and impassioned plea to his fellow botanists in 1966 for protection of the wild world.[45] Iltus claimed that:

> Technology has promised us a post-evolutionary heaven in which wild nature has a very minor role. Molecular biology, too, has gleeful visions of genetic manipulations of DNA which would change the face of all creation and recast man into a "perfect" image. Others dream about a cheer-

> ful if dull world with unlimited opportunity for at least 40 billion people, as a recent issue of *Soviet Life* pointed out in all seriousness.... [But] life's diversity is threatened with imminent destruction....[It] will be all but over in 20 or 30 years for this exuberant biotic wealth.
>
> The crisis for all the living is here and now. The world of the future promises to be flower-less, animal-less, and lifeless, except for masses of people.... Surely [the graying heads of the National Academy of Sciences] in their wisdom, must see that the living world is falling apart at the seams.... One more generation of ecological idiots is one too many.[46]

Iltus also argued for protection of the remaining wildness on Earth on the basis of the "genetic relationship" between humans and wild nature. He proposed a combination of the wild with the best that civilization has to offer, claiming that the human species needs contact with "the natural environment in which it evolved and for which it is genetically programmed":

> We, who are Darwin's children, can thus easily appreciate that like the need for love, the need for nature, the need for its diversity and beauty, has a genetic basis.... I have tried to show that separated from nature the human animal as a biological unit is in most ways a meaningless bundle of adaptations.... To understand our very existence, we need wilderness to know where we, the human species, came from. Yet we are rapidly becoming cultural and evolutionary orphans—a people without a past, a species out of context.

Iltus describes how mental institutions have used contact with Nature to help rehabilitate patients with long-term acute mental illness. And he reminds us that "Edgar Anderson and Carl Sauer have long pointed out that we have much to learn from primitive people and their ways."[47]

In *Nature and Madness,* Shepard traces his earlier intellectual journey. In *Man in the Landscape,* he examined the origins of an "environmental esthetic" but found that "too fragile and wayward" to provide a basis for ecological protection. He says that:

> Beginning again, in *The Tender Carnivore and the Sacred Game,* I sought a more durable model for ourselves in the deep history of ancestral hunters-gatherers. In an unexpected way I was rewarded, for foraging is a more thoughtful activity than I knew. I felt that I had glimpsed a central figure of consciousness, whose expressions in intelligence and speech appeared to be bound in each individual human emergence, as well as in that of the whole species, to plants and animals. I attempted to trace that emergence in *Thinking Animals.* Meanwhile, the sense of a normative psychogenesis became a compelling idea, of development as an organ.

Nature and Madness became an examination of "general, culturally-ratified distortions of childhood, of massive disablement of ontogeny as the basis of irrational and self-destructive attitudes toward the natural environment...[and a description of] where I think such a healthy ontogeny should lead."[48]

Drawing upon the human developmental psychology of Erik Erickson, Eric Fromm, Joseph Pearce, Edith Jacobson, and others, Shepard combines this with anthropological studies of primal peoples to develop a model of normal human ontogeny. In contrast with the rituals and rites of passage of primal peoples, young people raised wholly in urban settings with no such rituals or contact with the wild face insurmountable ontogenetic problems. He says, "Let us suppose, with some evidence, that the city is typically a sink of psychological problems. In the individual these are partly caused by city life, but in the long view they cause the city." Industrial urban consumer society, with its accelerating destruction of Nature, "works" because people are "ontogenetically stuck": they remain in an adolescent phase of development.[49] In summary, Shepard suggests that

> perhaps we do not need new religious, economic, technological, ideological, esthetic, or philosophical revolutions. We may not need to start at the top and uproot political systems, turn life-ways on their heads, emulate hunters and gatherers or naturalists, or try to live lives of austere privation or tribal organization. The civilized ways inconsistent with human maturity will themselves wither in a world where children move normally through their ontogeny.[50]

At the Wilderness and Civilization conference held in Estes Park, Colorado in 1989, I saw Paul Shepard again. Although we had corresponded off and on, I had not seen Paul for fifteen years, since the Claremont conference. This time there was an opportunity for a long uninterrupted chat with Paul and his wife Flo, which included a story about a bear hunt he had taken part in, and his concern that his mentor, Paul Sears (then in his 90s and living in New Mexico) be properly appreciated and remembered. As he lectured on his paper written for the conference ("A Post-Historic Primitivism"), I remember being impressed by the tremendous clarity and power of his presentation.

"A Post-Historic Primitivism" is a fine summary and overview of most of his work, which reviews and up-dates the literature on hunter/gatherers.[51] He points out that "almost everything that the layman generally thought to be true of [hunter/gatherers] was wrong." He discusses the cultural and psychological problems arising from the beginnings of agriculture, taking on Wendell Berry, and various ecofeminist, animal rights, and vegetarian critics in the process. He even has a few words of criticism of Gary Snyder while acknowledging the underlying similarities of their positions. While agreeing that we can't literally "go back" to ancient hunting/gathering ways of life, culture, he claims, is nevertheless a "mosaic." This means that certain key features of various ways of life can be transported from culture to culture and reincorporated. He concludes by laying out the main features of a post-modern cynegetic culture which he refers to as a Paleolithic counter-revolution or "Pleistocene paradigm."

In the most recent paper of Shepard's I have read ("Wilderness is Where My Genome Lives"), he speculates that the decoding of the human genome may finally and pragmatically result in the awareness of the universality of our genetic human nature: "that we are Pleistocene hominids keyed with infinite exactitude to the savannas and other ecotones of the earth wilderness."[52] In the best traditions of modern deconstructionism, Shepard has in effect "deconstructed" the ideologies of modernism, but avoided deconstructionism's total relativism by arriving at the rock-bottom reality of Pleistocene genetic human nature. Zen Buddhism can also be seen as an early attempt at "deconstructionism": rejecting ideologies and theoretical constructions in favor of "direct unmediated experience." For Zen Buddhism, the spiritual task is one of attaining human freedom by peeling back the layers of social conditioning to arrive at one's "original nature": again a rock bottom reality or datum for our species which Snyder also understands as Pleistocene human nature. As Snyder puts it, "People and places that will not be managed—they are called "wild"..."Self-thus," Chinese word for nature—not programmed—generating its own rules from within."

There are those who see in Spinoza a parallel with Zen Buddhism (despite Spinoza's metaphysical rational metaphysical system, which is—like Wittgenstein's metaphysics in the *Tractatus*—a rational guide to the interrelatedness of reality and human psychology, but designed ultimately to lead to human freedom through "intuitive knowing"). Spinoza's metaphysics and psychology of human maturity, together with intimate bonding with wild Nature, has provided Arne Naess with a path to the same kind of universal human nature—the "ecological self."[53]

Post-modern deconstructionists such as Donna Haraway and Jim Cheney argue against the possibility of universals, and for total cultural diversity and "multiculturalism." Although Cheney favors a bioregional appropriate technology approach, Haraway appears to take an uncritical urban stance towards progress and various forms of megatechnology. Some ecofeminists, such as Irene Javors and Carol Sanchez, go so far as to uncritically endorse urban industrial society and its technology: industrial urbanism, they claim, should be sacralized (industrial cities, petrochemicals and fossil fuels, television, computers, and even nuclear power!). Freya Mathews has suggested that "perhaps here, in the heart of the metropolis, Nature is at its wildest."[54] Surely, this is a play on words.

Perhaps these ecofeminists have never read Roszak's *Person/Planet* (1978), and/or are unmoved by, for example, Heidegger's crushing indictment of the negative effect (on both humans and non-humans) of the modern technological disclosure of Being; what Heidegger calls "the question concerning technology." These writers call themselves ecological feminists, but one begins to wonder just what ecological literature they have been reading!

For Shepard and Snyder, genetic Pleistocene human nature, together with the naturally evolving evolutionary processes of wild plants, animals and ecosystems co-evolving over hundreds of thousands of years without radical human intervention, are the universals against which human cultures—and cultural goals, activities, ideals of human personality types, and standards of sanity—must be measured. Tremendous diversity of individuals and cultures existed among primal peoples, but this diversity consisted of variations based upon these universals.

Shepard's analysis of the "ecopsychology of the wild," and his discussions of the "madness" (the "sickness") of industrial society, relies in part on Eric Fromm's concept of "the sane society." In a similar fashion, Vice-President Al Gore uses the psychological parallel of the "dysfunctional family" in characterizing modern industrial society as a "dysfunctional civilization" (with its psychopathology of consumption addiction).[55] But in comparison with the analyses of Shepard and Snyder, Gore obviously only begins to scratch the surface of these issues.

The historian Lewis Mumford is the great early critic of the myths of Western progress, the industrial society, megatechnology, and the "megamachine."[56] Shepard and Snyder push this critique back to the universality of the Paleolithic mind, and, in the case of Shepard, to the reality of a normal human ontogeny. It seems clear that, in following the lead of Thoreau and Muir, Snyder and Shepard are right in taking the wild/domestic issue as the central environmental, ecological, and ecophilosophical concern.

In contrast with Shepard and Snyder, Teilhard de Chardin, Murray Bookchin (and others influenced by the German idealist Hegelian-Marxist and modern social-science tradition) are anthropocentrically fixated on the human "social construction of reality": they claim that humans have created a "second nature" consisting of civilization, which has wholly transcended, and is separate from, the "first world" of wild nature and primal peoples.[57] Modern humans living in "second nature" (the image here is of a wholly subjectivistic human urban-technological realm free-floating in space above the rest of Nature) are thereby not subject to the biological laws of ecology that apply to "first nature." For example, the size of the human population, for most of the proponents of the "second nature" concept, can apparently increase indefinitely since it is not subject to ecological constraints. The modern task for Bookchin is to "rationally" incorporate wild "first nature" (which Bookchin thinks is determined and "unfree") into humanized civilized "second nature," with humans guiding the evolutionary processes, thereby creating a synthesis of the two in what he calls "free nature."

A sharp contrast offers itself here: Bookchin and others in the Enlightenment Hegelian/Marxist tradition (and also in the traditional Christian tradition, for that matter) think that wild nature and primal peoples are unfree; Thoreau, Muir, Snyder, Shepard, and Naess, think that the wild is free, and that the contemporary direction of civilization is making both humans and Nature increasingly unfree. Bookchin denies, or is unaware of, the universality of the Paleolithic mind; Snyder and Shepard hold that humans will be free only when we again attain and actualize the Paleolithic mind and scale-down human habitation on the Earth to fit in with the spontaneous direction of wild evolutionary processes. As one commentator has remarked, "Snyder's major accomplishment...is a rediscovery and reaffirmation of wilderness....[He] attempts to push it eastward, to reverse America's historical process, to urge the wilderness to grow back into civilization. . . ."[58]

The alternative to "wildness as the preservation of the world" is Baudrillard's characterization of "hyper-reality": a Disneyland megatechnological artificial world of immense shopping malls, human overpopulation, totalitarian societies, seriously atrophied and diminished human beings, and rapidly escalating ecological and environmental deterioration. Those scientists, technologists, corporate and business leaders, and post-modern intellectuals who advocate, or acquiesce in, these megatechnological artificial visions are, as Shepard points out, urban minds who are "ontogenetically stuck" and thereby alienated from the wild.[59] The corporate industrial megamachine is now pushing towards total commercialization and control over the world, thereby further warping the human psyche. The present rate of destruction of the wild can not wait until society changes to the point where humans can, once again, proceed through their normal ontogeny.

The theologian Thomas Berry, along with the world's leading professional ecologists such as Paul Ehrlich and E. O. Wilson, are acutely aware of the enormous magnitude of the environmental crisis and the destruction of the wild. Berry has remarked that people "just don't get it. They don't comprehend how deeply rooted it is, the crisis that confronts us!....The order of magnitude of the present catastrophic situation is...so enormous, so widespread, and we don't know what we are doing." Berry has emphatically pointed out that the basic conflict in modern society is between the visions of the industrial entrepreneur and the visions of the ecologist: this is "both the central human issue and the central earth issue of this late twentieth-century."[60]

More recently, Berry has claimed that

> reconciliation between [the developers and the ecologists] is especially difficult because the commercial-industrial powers have so overwhelmed the natural world in these past two centuries that there is, to the ecologist, no question of further adaptation of natural systems to the human. The oppression of the natural world by the plundering of the industrial powers has so endangered the basic functioning of natural forces that we are already on the verge of a total dysfunctioning of the planet. We cannot mediate the situation as though there were presently some minimal balance already existing that could be slightly modified so that a general balance could come into being. The violence already done to the Earth is on a scale beyond all understanding.... The change required by the ecologist is a drastic reduction in the plundering processes of the commercial industrial economy....Never before has the human community been confronted with a situation that required such a sudden and total change in life style under the threat of a comprehensive degradation of the plane....A special poignancy is experienced in the realization that future generations will be living amid the ruins; not simply amid the ruined infrastructures of the industrial world, but amid the ruins of the natural world itself.[61]

Worldwatch Institute researcher Sandra Postel, after outlining the current ecological state of the world, claims that most modern humans are in a state of psychological denial concerning the magnitude and seriousness of the crisis. Unfortunately, she claims, the 1990s are the "decisive decade" to seriously turn things around.[62]

In 1973, Paul Shepard claimed that "It is impossible to overestimate the ecological crime of species extinction.... Extinction is caused by alteration of the habitat. The measures necessary to avoid it are the same that preserve the biosphere as a whole. The prevention of extinction should be the criterion for a plan or policy of environmental activity of any kind."[63] Most contemporary ecologists and conservation biologists agree with Shepard. For instance, after running through an analysis of our contemporary ecological ills (human overpopulation, greenhouse effect, ozone depletion, acid rain, etc. etc.) Paul and Anne Ehrlich claim that "the ravaging of biodiversity, in our view, is the most serious single environmental peril facing civilization." The main solution to the environmental crisis, they claim, is to "reduce the scale of the human enterprise." E. O. Wilson estimates that human-caused species extinction has accelerated from approximately 1000 each year throughout the 1970s to more than 10,000 species per year at present.[64]

Meanwhile, the industrial elites of the world, through the United Nations Brundtland Report and the Rio environmental conference, relentlessly push for continued growth and development (and further destruction of habitat) under the guise of "sustainable development" and global technological management schemes for the "resources" of the biosphere, thereby trying to outflank the concerns of ecologists.[65]

It is heartening to see environmental ethics and ecophilosophy theorists (and academic intellectuals across a wide spectrum of disciplines) now beginning to take seriously the question of the propriety (the incredible negative ecological and human impact) of modern industrial societies. This involves such issues as the sacred versus the profane, the domination of economics, international markets, and corporations in contemporary society (including the accelerating take-over of schools and universities by corporations), consumption as a way of life and ubiquitous demand-creating advertising, the further colonization and exploitation of the Third and Fourth Worlds by industrial societies, and the corporate and scientific/technology establishment's visions of artificial mega-technology futures in which wildness has been obliterated. As Paul Shepard pointed out decades ago, the ideological status of ecology is that of a resistance movement. Ecophilosophy and the world-wide environmental movement must now function as a resistance movement to the establishment of a totally controlled international corporate world, megatechnology futures, and further destruction of the wild, as transition strategies are developed to attain post-modern ecologically sustainable societies in which Paleolithic/ecological consciousness prevails.

Notes

1. See, e.g., David Suzuki and Peter Knudtson, *Wisdom of the Elders* (New York: Bantam, 1992).

2. See Frank F. Darling and John Milton, eds., *Future Environments of North America* (Washington, D. C.: Natural History Press, 1966; Anne Chisholm, *Philosophers of the Earth: Conversations with Ecologists* (New York: Dutton, 1972). Recent ecocentric interpretations of Thoreau, Muir, Leopold, Snyder, *et al.* are found in Max Oelschlaeger, *The Idea of Wilderness: From Prehistory to the Age of Ecology* (New Haven: Yale University Press, 1991).
3. For discussions of the Claremont ecophilosophers and the early development of ecophilosophy, see George Sessions, "Shallow and Deep Ecology: A Review of the Philosophical Literature," in Robert Schultz and J. Donald Hughes, eds., *Ecological Consciousness: Essays from the Earthday X Colloquium*, University of Denver, April 21-24, 1980 (Washington, D. C.: University Press of America, 1981, 391-462; George Sessions, "The Deep Ecology Movement: A Review," *Environmental Review* 11, 2 (1987):105-25.
4. Press coverage of the Theology of Survival Conference appeared in the *San Francisco Examiner-Chronicle*, "This World," May 10, 1970, and *Time*, June 8, 1970, 49.
5. See John Cobb, Jr., "The Population Explosion and the Rights of the Subhuman World" IDOC-North America, Inc. (Sept. 12, 1970) 40-60; reprinted in R. T. Roelofs, J. N. Crowley, et al., eds., *Environment and Society* (Englewood, N. J.: Prentice-Hall, 1974). See also John Cobb, Jr., *Is It Too Late?: A Theology of Ecology* (Bruce, 1972); Charles Birch and John Cobb, Jr., *The Liberation of Life* (Cambridge: Cambridge University Press, 1985); Herman Daly and John Cobb, Jr., *For the Common Good* (Boston: Beacon Press, 1989).
6. The "Philosophy and Environmental Crisis" papers were published three years later: William Blackstone, Jr., ed., *Philosophy and Environmental Crisis* (Athens: University of Georgia Press, 1974).
7. See Pete A. Y. Gunter, *The Big Thicket: A Challenge for Conservation* (Austin, Tex.: Jenkins, 1973); Gunter, "The Perils of Conservationist Politics: Life in the Trenches," in Max Oelschlaeger, ed., *After Earth Day: Continuing the Conservation Effort* (Denton: University of North Texas Press, 1992), 28-34; Gunter is discussed in George Sessions, ed., *Ecophilosophy II* (1979):4-5.
8. See e.g., Garrett Hardin, "The Tragedy of the Commons," *Science* 162 (1968); G. Hardin and John Baden, eds., *Managing the Commons* (New York: Freeman, 1977); William Leiss, *The Domination of Nature* (New York: Braziller, 1972); Leiss, *The Limits of Satisfaction* (Toronto: University of Toronto Press, 1976); Roderick Nash, *Wilderness and the American Mind* (New Haven: Yale University Press, 1967); John Lilly, *The Mind of the*

Dolphin (New York: Doubleday, 1967); Vine Deloria, Jr., *God is Red* (New York: Grossett, 1973); John Livingston, *One Cosmic Instant* (New York: Macmillan, 1973); Joseph Meeker, *The Comedy of Survival* (New York: Scribner's, 1974); Gary Snyder, "The Yogin and the Philosopher," in Snyder, *The Old Ways* (San Francisco: City Lights Books, 1977), 9-14; Arne Naess, "The Shallow and the Deep, Long-Range Ecology Movements: A Summary," *Inquiry* (Oslo) 16 (1973):95-100, reprinted in P. Struhl and K. Struhl, eds., *Philosophy Now,* 2nd ed. (New York: Random House, 1975), 599-604; for a discussion of the early development of Naess's ecosophy, see George Sessions, "Deep Ecology as a World View," in M. Tucker and J. Grim, eds., *World Views and Ecology* (Lewisburg, Penn.: Bucknell University Press, 1993), 207-27.

9. Much of the material in this ms. was published as "Anthropocentrism and the Environmental Crisis," *Humboldt Journal of Social Relations* 2, 1 (1974):71-81.
10. My paper "Panpsychism vs. Modern Materialism: Some Implications for an Ecological Ethic" (1974), along with the Hartshorne/Birch/Cobb papers, were never published. For a discussion of these events, see Warwick Fox, *Towards a Transpersonal Ecology* (Boston: Shambala, 1990), 60-69. For a recent attempt to reconcile Whiteheadian ethical extensionism with biocentric egalitarianism, see David Ray Griffin, "Whitehead's Deeply Ecological World View," in Tucker and Grim, *World Views and Ecology*, 190-206.
11. John Rodman tried to find a publisher for the collected conference papers, but to no avail. As environmental editor of the *North American Review,* Meeker had a few of the papers published there. See the papers by Hardin, Shepard, and Nash in the *North American Review* 259 (1974).
12. For a discussion of the events surrounding the establishment of the *New Natural Philosophy* program, together with Arne Naess's visit to Claremont in 1977, see Dolores LaChapelle, *Sacred Land Sacred Sex: Rapture of the Deep* (Durango, Col.: Kivaki Press, 1988), 11-12.
13. John Rodman, "The Dolphin Papers," *North American Review* 259 (1974):3-26; see also Rodman, "Paradigm Change in Political Science," *American Behavioral Scientist* 24 (1980):49-78; Rodman's series of papers was discussed extensively in my *Ecophilosophy II* newsletter (May 1979).
14. John Rodman, "Four Forms of Ecological Consciousness, Part One: Resource Conservation," presented at the annual meeting of the American Political Science Association, 1976, unpublished ms.; Rodman, "The Liberation of Nature?" *Inquiry* (Oslo) 20 (1977):83-131.

15. John Rodman, "Ecological Resistance: John Stuart Mill and the Case of the Kentish Orchid," presented at the annual meeting of the American Political Science Association, 1977, unpublished ms.; see also Rodman "Theory and Practice in the Environmental Movement," in *The Search for Absolute Values in a Changing World*, Vol. 1 (New York: The International Cultural Foundation, 1978).
16. Published as John Rodman,"Four Forms of Ecological Consciousness Reconsidered," in Thomas Attig and Donald Scherer, eds., *Ethics and the Environment* (Englewood Cliffs, N. J.: Prentice-Hall, 1983), 82-92.
17. Richard Routley, "Is There a Need for a New, An Environmental Ethic?" reprinted in Michael Zimmerman, J. Baird Callicott, *et al.*, eds., *Environmental Philosophy: From Animal Rights to Radical Ecology* (Englewood Cliffs, N.J.: Prentice-Hall, 1993), 12-21.
18. See e.g., J. Baird Callicott, "Elements of an Environmental Ethic," *Environmental Ethics* 1 (1979):71-81; Callicott, "Hume's Is/Ought Dichotomy and the Relation of Ecology to Leopold's Land Ethic," *Environmental Ethics* 4 (1982):163-74; Holmes Rolston, III, "Can and Ought We to Follow Nature," *Environmental Ethics* 1 (1979):7-30; Rolston, "Values in Nature," *Environmental Ethics* 3 (1981):113-28; Callicott, "Animal Liberation: A Triangular Affair," *Environmental Ethics* 2 (1980):311-38.
19. For a discussion of these issues, see, for instance, Michael Zimmerman, "General Introduction" in Zimmerman, *Environmental Philosophy*, v-x.
20. Louis Pojman, ed., *Environmental Ethics* (Boston: Jones and Bartlett, 1994), xiii, 64.
21. See the editorial by Eugene Hargrove, "New Directions" *Environmental Ethics* 3 (1981):291-2.
22. Rodman, "The Liberation of Nature?":98, 103.
23. Rodman, "Four Forms of Ecological Consciousness Reconsidered":88-92.
24. See, for instance, Stephen Fox, *John Muir and His Legacy: The American Conservation Movement* (Boston: Little, Brown, 1981); Michael P. Cohen, *The Pathless Way: John Muir and American Wilderness* (Madison: University of Wisconsin Press, 1984); Oelschlaeger, *Idea of Wilderness*, 133-204.
25. See Michael P. Cohen, *The History of the Sierra Club: 1892-1970* (San Francisco: Sierra Club Books, 1988).
26. Some of the issues I am raising here are discussed in David Strong, "Disclosive Discourse, Ecology, and Technology," *Environmental Ethics* 16, 1 (1994):89-102.
27. Oelschlaeger, *Idea of Wilderness*, 1-30, 320-53; see also Jerry Mander, *In the*

Absence of the Sacred: The Failure of Technology and the Survival of the Indian Nations (San Francisco: Sierra Club Books, 1991).

28. Paul Shepard, *Man in the Landscape: A Historic View of the Esthetics of Nature* (New York: Knopf, 1967).
29. Paul Shepard and Daniel McKinley, eds., *The Subversive Science: Essays Toward an Ecology of Man* (Boston: Houghton Mifflin, 1969); Paul Shepard and Daniel McKinley, eds., *Environ/Mental: Essays on the Planet as a Home* (Boston: Houghton Mifflin, 1971).
30. See, for example, Frank Egler, *The Way of Science: A Philosophy of Ecology* (New York: Hafner, 1970), 137; David Barash,"The Ecologist as Zen Master," *American Midland Naturalist* 89 (1973):214-17; Hwa Jol Jung, "The Splendor of the Wild: Zen and Aldo Leopold," *Atlantic Naturalist* 29 (1974).
31. Shepard, "Ecology and Man," 8
32. Ibid., 4-5.
33. Ibid., 9; Paul Sears, "Ecology—a Subversive Subject," *Bioscience* 14, 7 (1964).
34. Paul Shepard, *The Tender Carnivore and the Sacred Game* (New York: Scribners, 1973).
35. Shepard, *Tender Carnivore,* 237.
36. Ibid., 260.
37. Ibid., 177, 233-73; see also the section on Shepard's *Tender Carnivore* in Bill Devall and George Sessions, *Deep Ecology* (Salt Lake City: Gibbs Smith, Inc., 1985), 172-76.
38. Paul Shepard, *Thinking Animals: Animals and the Development of Human Intelligence* (New York: Viking Press, 1978).
39. Paul Shepard & Barry Sanders, *The Sacred Paw: The Bear in Nature, Myth, and Literature* (New York: Viking Penguin, 1985); Paul Shepard, "To Friends of The Sacred Paw," Letter dated April 12, 1993.
40. Shepard, *Thinking Animals,* 246-52; see also the discussion of Shepard's *Thinking Animals* in George Sessions, "Ecocentrism, Wilderness and Global Ecosystem Protection," in Max Oelschlaeger, ed., *The Wilderness Condition: Essays on Environment and Civilization* (San Francisco: Sierra Club Books, 1992), 97-98.
41. Theodore Roszak, *The Voice of the Earth* (New York: Simon and Schuster, 1992), 13, 320.
42. Warwick Fox, (Boston, Shambala, 1990). Arne Naess sees Fox's transpersonal ecology, as well as his own concept of the "ecological self" and "Ecosophy T," as existing at Level 1 of the Apron Diagram. The Deep Ecology

movement is best characterized by the deep questioning process and the Level II Deep Ecology platform. Andrew McLaughlin also argues that the Deep Ecology movement should be understood in terms of the Deep Ecology platform; see McLaughlin, *Regarding Nature: Industrialism and Deep Ecology* (New York: State University of New York Press, 1993), 172-90; McLaughlin, "The Heart of Deep Ecology," in George Sessions, ed., *Deep Ecology for the 21st Century* (Boston: Shambhala, 1994).

43. See Gary Snyder, *Turtle Island* (New York: New Directions, 1974); Snyder, "Good Wild Sacred," in Snyder, *Coming into the Watershed* (New York: Pantheon, 1994); Snyder, *The Practice of the Wild* (San Francisco: North Point Press, 1990); Arne Naess, "Self-Realization: An Ecological Approach to Being in the World," *The Trumpeter* 4, 3 (1987):35-42 (to appear in Sessions, *Deep Ecology for the 21st Century*).
44. For an important survey of Western culture from this perspective, see Oelschlaeger, *Idea of Wilderness*.
45. Hugh H. Iltus, "Whose Fight is the Fight for Nature?" *Sierra Club Bulletin* 52, 9 (October 1967):34-3; for other arguments that humans are genetically programmed for wild environments, see E. O. Wilson, *Biophilia* (Cambridge: Harvard University Press, 1984); Stephen Kellert and E. O. Wilson, eds., *The Biophilia Hypothesis* (Washington, D. C.: Island Press, 1993).
46. Ibid., 35, 37, 38.
47. Ibid., 35-6.
48. Paul Shepard, *Nature and Madness* (San Francisco: Sierra Club Books, 1982), ix.
49. Ibid., 93, 120.
50. Ibid., 128-29.
51. Paul Shepard, "A Post-Historic Primitivism," in Oelschlaeger, ed., *The Wilderness Condition*, 40-89.
52. Paul Shepard, "Wilderness is Where My Genome Lives," paper delivered at the 5th World Wilderness Congress, Tromsø, Norway, 1993.
53. For a discussion of Wittgenstein, Spinoza, Zen Buddhism, and direct intuitive knowledge and experience which goes beyond rational metaphysical systems, see George Sessions, "Western Process Metaphysics: Heraclitus, Whitehead, and Spinoza," in Devall and Sessions, *Deep Ecology*, 236-42.
54. See, e. g., Donna Haraway, *Simians, Cyborgs, and Women: The Reinvention of Nature* (London: Free Association Books, 1991); Jim Cheney, "Postmodern Environmental Ethics: Ethics as Bioregional Narrative," *Environmental Ethics* 11 (1989):117-34; Irene Javors, "Goddess in the Metropolis: Reflections on the Sacred in the Urban Setting," in Irene Diamond and Gloria Orenstein,

eds., *Reweaving the World: The Emergence of Ecofeminism* (San Francisco: Sierra Club Books, 1990; Carol Sanchez, "New World Tribal Communities," in J. Plaskow and C. Christ, eds., *Weaving the Visions* (New York: Harper and Row, 1989); Freya Mathews, "Relating to Nature," (1993), unpublished ms.

These views glorifying urban-industrial megatechnological society should be contrasted with those of bioregionalists who argue for scaled down, appropriate technology communities, and with the views of Helena Norberg-Hodge who documents the on-going destruction of the Ladakhi culture as the result of Western development, and "progress" through the introduction of market economies and high technology. She holds out traditional Ladakhi culture as a social model for the de-development of the West. See Helena Norberg-Hodge, *Ancient Futures: Learning from Ladakh* (San Francisco: Sierra Club Books, 1991); see also Vandana Shiva, *Monocultures of the Mind: Biodiversity, Biotechnology and the Third World* (Malaysia: Third World Network, 1993).

55. See Shepard, *Nature and Madness*, xi, 4-5; Al Gore, Jr., *Earth in the Balance: Ecology and the Human Spirit* (Boston: Houghton Mifflin, 1992), 216-37; see also Greta Gaard, "Review of Gore, Earth in the Balance," *Environmental Ethics* 15, 4 (1993):363-69.
56. In addition to Mumford's writings, see Theodore Roszak, *Where the Wasteland Ends: Politics and Transcendence in Postindustrial Society* (New York: Doubleday, 1972), where Roszak builds upon Mumford's critique.
57. For a clear description of Bookchin's views, see Janet Biehl, "Dialectics in the Ethics of Social Ecology," reprinted in Zimmerman, *Environmental Philosophy*, 374-89; for interesting ecological/anthropological commentary on this "two worlds picture" which is in accord with the orientation of Shepard and Snyder, see Loren Eiseley, "The Last Magician," in Eiseley, *The Invisible Pyramid* (New York: Scribner's, 1970), 137-56.
58. L. Edwin Folsom, "Gary Snyder's Descent to Turtle Island: Searching for Fossil Love," *Western American Literature* 15 (1980):103-21.
59. For current discussions of these issues, see Thomas Birch, "The Incarceration of Wildness: Wilderness Areas as Prisons," *Environmental Ethics* 12, 1 (1990):3-26; Mander, *In the Absence of the Sacred*; Jeremy Rifkin, *Biosphere Politics* (New York: Crown Publishers, 1991); Ivan Illich, *In the Mirror of the Past* (New York: Marion Boyers, 1992); Alan Durning, *How Much is Enough?: The Consumer Society and the Future of the Earth* (New York: Norton, 1992).
60. Berry is quoted in Marjorie Hope and James Young, "A Prophetic Voice:

Thomas Berry," *The Trumpeter* 11, 1 (1994):2-9; Thomas Berry, "The Viable Human," reprinted in M. Zimmerman, *Environmental Philosophy*, 171-81.

61. Thomas Berry, "The New Political Alignment," (1993), unpublished ms.
62. Sandra Postel, "Denial in the Decisive Decade," in Lester Brown, ed., *State of the World* (New York: Norton, 1992), 3-8.
63. Shepard, *Tender Carnivore*, 266.
64. Paul and Anne Ehrlich, *Healing the Planet: Strategies for Resolving the Environmental Crisis* (Reading, Mass.: Addison-Wesley, 1991), 35, 37; E. O. Wilson, *The Diversity of Life* (Cambridge: Harvard University Press, 1992) p. 351; see also G. Tyler Miller, *Living in the Environment* 8th edition (Belmont, Cal.: Wadsworth, 1994).
65. For an excellent critique of the Rio conference, "sustainable development," global management schemes, and Third World environmental concerns, see Wolfgang Sachs, "Global Ecology and the Shadow of 'Development,'" in Sachs, ed., *Global Ecology* (New Jersey: Zed Books, 1993), 3-21; for another excellent discussion of Third World environmental problems, see Paul Harrison, *The Third Revolution: Environment, Population and a Sustainable World* (London: I. B. Tauris, 1992).

Paul Shepard and the Development of Mid-Twentieth-Century Ecological Metaphysics

J. Baird Callicott

THE FRENCH PHRASE, *FIN-DE-SIÈCLE* ("END OF CENTURY") has been incorporated into English to name the mood pervading Western civilization one hundred years ago—at the end of the nineteenth century.[1] The 1890s was a decade characterized by "sophistication, world-weariness, and fashionable despair." The affective dimension of the Western Zeitgeist seems a centennial cycle, since the 1990s is definitely marked by a similar state of mind. It is manifest in ecology no less than in other domains of culture—so much so, indeed, that one might speak of a *fin-de-vingtième-siècle* ecology. Ecology at the end of the twentieth century is certainly mathematically sophisticated; ecologists are weary of hearing about the "balance of nature" and the dire necessity to preserve it; and they despair of finding any unifying theoretical paradigm.

During the mid-twentieth century, however, ecology was unified by the ecosystem paradigm. It was proposed by Arthur G. Tansley in 1935; operationalized by Raymond L. Lindeman in 1942; and institutionalized by Eugene P. Odum during the 1950s, '60s, and '70s in the several editions of his *Fundamentals of Ecology*. Tansley had introduced the nonmetaphorical concept of an ecosystem deliberately to obviate the scientifically malodorous holism suggested by the supraorganism paradigm championed by early twentieth- century ecology's dean, Frederick Clements. For a model, Tansley looked to the hard, reductive science of physics. The defining features of the ecosystem paradigm were first, a focus on quantifiable energy capture, and second, adding abiotic factors, such as the mineral components of soil, to relationships among organisms, as the proper subject of ecological study. But Odum, ecology's dean during the mid-twentieth century, ironically, portrayed the ecosystem paradigm to be no less holistic than Clements's supraorganismic paradigm: "By a system," he writes, "we mean just what *Webster's Collegiate Dictionary* defines as 'regularly interacting and interdependent components forming a unified whole.'"[2] Odum understood terrestrial nature to be characterized by hierarchically ordered multiple "levels of organization"—molecules, genes, cells, organs, organisms, populations, communities, ecosystems. "The largest and most nearly self-sufficient system we know about is often designated as the biosphere or ecosphere, which includes all of the earth's living organisms interacting with the physical environment as a whole so as to maintain a steady-state system intermediate in the flow of energy between the high energy input of the sun and the thermal sink of space."[3] At mid-century it occurred to a few insightful thinkers that—unlike some other subdisciplines of biology, for example, physiology or systematics—ecology had profound philosophical implications. The then "new ecology," in tandem with the "new physics," was giving birth to a brand new world view. One such thinker was Aldo Leopold. Another was Paul Shepard. In 1939, only four years after Tansley's programmatic paper and three years before Lindeman's pioneering quantificational study of energy flows through the trophic pyramid of a well-defined aquatic ecosystem, Leopold characterized the terrestrial world represented in ecosystem ecology as a

> fountain of energy flowing through a circuit of soils, plants, and animals. Food chains are the living channels which conduct energy upward....The upward flow of energy depends on the complex structure of the plant and animal community, much as the upward flow of sap in a tree depends on its complex cellular organization. Without this complexity normal circulation would not occur. Structure means the characteristic numbers, as well as the characteristic kinds and functions of the species.[4]

At the dawn of Western philosophy, Protagoras famously remarked that "Man is the measure of all things: of existing things that they exist; of nonexistent things, that they do not exist." Each of us human beings enjoys a personal identity, a self, by which we measure existing and nonexistent things. We appear to ourselves to be independent, free-standing entities among other similar entities—other human beings, other animals (such as deer and bears) other living organisms (such as trees and bushes), and non-living things (such as boulders). "The world" appears to be a collection, an aggregate, of such things, animate and inanimate, living and non-living. And the relationships between the substantial things appear to be correspondingly insubstantial; they belong in the realm of the nonexistent. But classic ecosystem ecology paints another picture. Relationships—"the complex structure of the plant and animal community," in Leopold's words—outlast the transient relata. And the fundamental reality is energy. Animals (we human beings included) and plants appear to be mere ephemeral conduits in the stable, patterned flow of solar energy from source to sink. At the zenith of mid-twentieth-century ecological holism, Harold Morowitz characterized the Weltanschauung emerging from ecology as follows:

> Viewed from the point of view of modern [ecology], each living thing is a dissipative structure, that is, it does not endure in and of itself but only as a result of the continual flow of energy in the system.... From this point of view, the reality of individuals is problematic because they do not exist per se but only as local perturbations in this universal energy flow....An example might be instructive. Consider a vortex in a stream of flowing

> water. The vortex is a structure made of an ever changing group of water molecules. It does not exist as an entity in the classical Western sense; it exists only because of the flow of water through the stream. If the flow ceases the vortex disappears. In the same sense the structures out of which the biological entities are made are transient, unstable entities with constantly changing molecules dependent on a constant flow of energy to maintain form and structure.[5]

A contemporary of Eugene Odum, Paul Shepard is an important architect of the metaphysical revolution portended by mid-twentieth-century ecology. In one of his earliest papers, "A Theory of the Value of Hunting," Shepard points out that death, which hunters deliberately inflict upon their game, is conventionally regarded as "the worst of natural events." Along with the more pedestrian biological reasons for more positively valuing death as a natural event—death is the complement of life, not its antithesis, etc.—he connects the conventional view to something more fundamental: the prevailing Western ontology. Antipathy toward death in general, Shepard writes, "is part of the naive assumption that order in nature is epitomized by living objects rather than complex flow patterns of which objects are temporary formations."[6] In this single sentence he anticipates Morowitz by nearly fifteen years. If Leopold's tree-sap analogy and Morowitz's whirlpool analogy characterize the emerging ontology of ecosystem ecology at mid-century, Shepard crafts a graphic analogy to characterize the obsolescing object-centered ontology of mainstream Western science and civilization: "It presumes naively that the landscape is a room-like collection of animate furniture."[7] And he hints at a similarity between the ontology of the then new ecology and the ontology of the new physics: Both are "best describable in terms of events which constitute a field pattern."[8]

In the same early essay, Shepard goes on to suggest that ecosystem ecology also points to a transformation in Western moral psychology which hunting can help to facilitate: "What do the hunt and kill actually do for the hunter? They confirm his continuity with the dynamic life of animal populations, his role in the complicated cycles of elements, in the sweep of evolution, and in the patterns of the flow of energy."[9] From the point of view of mid-twentieth-century ecosystem ecology, one's self and the surrounding world are continuous. To be sure, one's self and the surrounding world are not indistinguishable, but neither can one conceive one's self to be entirely separate and distinct from the surrounding world.

Shepard expressed the philosophical implications of mid-twentieth-century ecosystem ecology only incipiently in "Toward a Theory of the Value of Hunting." A decade later, he articulated them much more fully and forthrightly in the introduction, entitled "Ecology and Man—a Viewpoint," to *The Subversive Science*, a collection of essays that he edited with Daniel McKinley. In that piece he concentrates less on the "field" ontology associated both with the new physics and mid-twentieth-century ecosystem ecology to develop what might be called a "relational" concept of the self.

First, Shepard describes the prevailing notion of the self in the Western tradition of thought as "an arrangement of organs, feelings, and thoughts—a Me—surrounded by a hard body boundary: skin, clothes, and insular habits."[10] Echoing Alan Watts, who had earlier caricatured the Western sense of self as "an ego in a sack of skin," Shepard suggests that "to see ourselves—or a plant or animal—as an isolated sack, a thing, a contained self" is encouraged by the very structure of Indo-European languages.[11] I'm not so sure what the structure of Western languages has to do with it, but certainly Western religion and philosophy have persistently identified the self with an immaterial and often immortal entity—a soul or psyche—that is imagined temporarily to inhabit our physical bodies, much as our physical bodies inhabit houses and cars. The self thus is both isolated from the surrounding physical world by its bodily cladding and, being an immaterial and otherworldly being, the self is also alienated from the surrounding physical world. According to this view, each self resides somewhere behind the eyes and between the ears of its physical vehicle peering fearfully out on an "other" and alien world.

In the West the self has been an individual in two senses of the word. The self is a psychic atom and thus individual in the sense of discrete or discontinuous. It is also individual in the sense of unique. No two selves are exactly alike. In the immensely influential Judeo-Christian tradition of thought, our unique personalities are given us by God. Thus we are what we are (immaterial sojourners in a strange land of physical objects) and we are who we are—the peculiar persons that we are, the unique individuals that we are—independently of any relationship with our social and natural milieus. According to this way of thinking, we are, in short, externally related to both to one another and to nature.

According to Shepard, "The alternative is a self as a center of organization, constantly drawing on and influencing the surroundings, whose skin and behavior are soft zones contacting the world instead of excluding it." Shepard's relational conception of the self implies that we human beings are physical beings, essentially, not psychic substances. To appropriate a term coined by the French phenomenologist Maurice Merleau-Ponty, we are "body-subjects," not unphysical souls uncomfortably inhabiting foreign bodies. From the ecological point of view—at least in 1969—we body-subjects are not discrete entities; we are, rather, continuous with our natural surroundings:

> Ecological thinking . . . requires a kind of vision across boundaries. The epidermis of the skin is ecologically like a pond surface or a forest soil, not a shell, so much as a delicate interpenetration. It reveals the self ennobled and extended rather than threatened as part of the landscape and the ecosystem, because the beauty and complexity of nature are continuous with ourselves.[12]

Then—expressly citing a chapter, "The World is Your Body," in Alan Watts's *The Book On the Taboo Against Knowing Who You Are*—Shepard goes on to an expansive conclusion: "we must affirm that the world is a being, a part of our own body."[13] The practical consequences of this enlarged relational conception of self are so obvious that Shepard hardly bothers to draw them in this essay, though he does so fully in *Nature and Madness*: environmental destruction is tantamount to self-destruction; biocide is suicide.

Images from some one other than Alan Watts recur in this essay of Shepard's. The briefly mentioned pond metaphor, to which Shepard subsequently returns for fuller illustration, is surely inspired by Raymond Lindeman's landmark article on ecosystem theory, "The Trophic-Dynamic Aspect of Ecology." Lindeman's measurements of "'energy availing' relationships" were carried out on a pond, Cedar Bog Lake, in Minnesota. Thus, for ecologists at mid-century, a pond was an emblematic ecosystem. Here is how Shepard elaborates the relational view of the self, taking the pond as a philosophical symbol:

> The pond is an example. Its ecology includes all events: the conversion of sunlight to food and the food-chains within and around it, man drinking, bathing, fishing, plowing the slopes of the watershed, drawing a picture of it, and formulating theories about the world based on what he sees in the pond. He and all the other organisms at and in the pond act upon one another, engage the earth and atmosphere, and are linked to other ponds by a network of connections like the threads of living protoplasm connecting cells in living tissues.[14]

Paul Shepard had, as I recall he once told me, a grant application turned down because his holistic ideas seemed to suggest, to the director of the foundation from which he had sought funding, "only followers, gray masses without the tradition of the individual."[15] Shepard seems concerned about that too and so insists that his proposed relational concept of the self is a complement to, not a wholesale substitute for, the tradition of the individual. "Both views are real," he writes, "and their reciprocity significant. We need them both to have a healthy social and human maturity."[16] Shepard himself stresses the relational view, since the ego-in-the-sack-of-skin view is constantly reiterated everywhere—in Sunday school, in civics classes, on television. It is "conferred on us," as Shepard puts it, "by the whole history of our civilization."[17]

But I don't see how we can consistently entertain them both. They are such radically and diametrically opposed conceptions of the self. Fortunately, the worry is baseless. An even more robust individuality, in the sense of personal uniqueness, is implied by the relational view. How so? Because no two individuals are constituted by the same set of relationships to their social and natural environments. Let me illustrate with another familiar image of mid-twentieth-century forensic ecology, the "web of life." Upon the relational view, each person—indeed, each individual organism—is, as it were, an intersection of the web's strands. Shepard, incidentally, condemns this image as "too meager and simple" since "a web is flat and finished."[18] And he is right to do so. But if we think of the web of life as multidimensional, the image is more apt. We are internally not externally related beings. That is, each of us is the unique person that he or she is, not because of some formally unique soul or psyche at the core of his or her being, but because he or she is a unique concatenation of extraordinarily rich, complex, and very real personal, social, and bioregional relationships. At one point, indeed, Shepard veritably exclaims "relationships of things are as real as the things."[19]

As I have elsewhere pointed out, the relational concept of self is endemic to Chinese thought, especially to the Confucian tradition. An individual is the son or daughter of a certain set of parents; a denizen of a certain village, town, or city; a butcher, baker, or candlestick maker; a resident of a certain bioregion. These are not roles external to a person, at which he or she plays; they are constitutive of the person himself or herself. If by some unfortunate circumstance they are taken away, the person's identity as an individual will have been destroyed, and a kind of ego collapse will ensue. A person who, whether consciously or not, conceives of himself or herself as internally related to his or her community and to his or her bioregion is the proverbial fish out of water when out of his or her social and environmental milieus. I should note that while the relational conception of self does not lack unique individuality, obviously it does lack "rugged individuality," independence, that is, from other people and from place.

If the relational self is, first and foremost, a body-subject, how can we account for the functions that were assigned to the soul or psyche in traditional Western religion and philosophy? Surely, the existence of our human minds, as Descartes argued, is proof positive that we are more than mere animals, that our animal bodies must be inhabited by an unphysical and specifically human "thinking thing." Shepard is impatient with this view: "The nonanimal forces," he writes,

> have developed a more subtle dictum to the effect that, "Man may be an animal, but he is more than an animal too!" The more is really what is important. This appealing aphorism is a kind of anesthetic. The truth is that we are ignorant of what it is like or what it means to be any other kind of creature than we are. If we are unable to define the animal's experience of life or "being an animal" how can we isolate our animal part?[20]

As to the allegedly species specific mental abilities of *Homo sapiens*, Shepard provides an ingenious relational theory of their origins, which he develops at great length in his book *Thinking Animals*. In "Ecology and Man—A Viewpoint" he argues, briefly, that "Internal complexity, as the mind of a primate, is an extension of natural complexity, measured by the variety of plants and animals and by the variety of nerve cells. The exuberance of kinds [is] the setting in which a good mind could evolve (to deal with a complex world)."[21] Thus we human beings owe our very intelligence—of which we are so proud and by which, ironically, we set ourselves apart from nature—to nature's mind-nurturing biological diversity and complexity.

Looking back at these two short early pieces by Paul Shepard leaves me feeling bitter-sweet. They are both great essays, but the fin-de-siècle climate in contemporary ecology leaves the philosophy they express without a solid scientific grounding. A field ontology plus a relational moral psychology may be an appealing viewpoint, but it can hardly, today, be called unequivocally ecological. In the new chapters of the recent second edition vingtième-fin-de-siècle of Nature's Economy, Donald Worster reports surveying British and American textbooks in ecology and finding that in all of them the ecosystem concept is muted and that in some it is not even mentioned. R. H. Whittaker's individualism has supplanted Odum's holism. In current fin-de-siècle ecology the existence of well-defined ecosystems is questionable; biotic communities are simply aggregates of plants and animals that have adapted to similar regimes on gradients of temperature, humidity, acidity, etc.; cooperation between community members is a myth (perhaps inspired by talk of flower-power in the '60s), though competition is real enough; perturbation not equilibrium is natural and normal; foundational concepts like "succession" and "climax" have been all but abandoned in favor of random change; and chaos mathematics is as useful in ecology as in climatology. So unsettled has become the state of ecological theory that Worster is reduced, at the end of his discussion, to reminding his readers, after all, that no organism or species of organism can survive without others.

The unhappy 1990s will soon be over and a new, perhaps more optimistic century, the beginning of a new millennium, will dawn. Hopefully, ecosystem ecology, will be successfully revived by contemporary architects of hierarchy theory such as R. V. O'Neil and T. F. H. Allen. One methodological problem with the classic ecosystem concept was how clearly to establish the boundaries of an ecosystem. Lindeman's pond is a convenient but misleading model, because its boundaries can be fairly easily and straightforwardly specified. Terrestrial and marine ecosystems cannot be so obviously delimited. Hierarchy theory claims to offer a rigorous, but flexible and relativized method for doing so in terms of temporal and spatial scale. If hierarchy theorists succeed in their bid to establish a unified paradigm for ecology, then a return to Paul Shepard's early essays on ecological metaphysics will be an unadulterated pleasure, indeed. Shepard's mid-twentieth-century ecological metaphysics will then be in need of revision, certainly, and development, but it could be a foundation upon which future environmental philosophers might construct a more mature ecological ontology and moral psychology.

Notes

1. Frederick C. Mish, ed., *Merriam-Webster's Collegiate Dictionary,* 10th ed. (Springfield, Mass.: Merriam-Webster, Inc. 1993), 436.
2. Eugene P. Odum, *Fundamentals of Ecology,* 3rd ed. (Philadelphia: W. B. Saunders, 1971), 4.
3. Ibid. 5.
4. Aldo Leopold, "A Biotic View of Land," *Journal of Forestry* 37 (1939), 728.
5. Harold Morowitz, "Biology as a Cosmological Science" *Main Currents in Modern Thought* 28 (1972), 156.
6. Paul Shepard, "A Theory of the Value of Hunting," *Transactions Twenty-fourth North American Wildlife Conference* (1959), 505.

7. Ibid.
8. Ibid., 506.
9. Ibid., 509.
10. Paul Shepard, "Introduction: Ecology and Man—A Viewpoint" in Paul Shepard and Daniel McKinley, eds., *The Subversive Science: Essays Toward an Ecology of Man* (Boston: Houghton Mifflin, 1969), 2.
11. Ibid.
12. Ibid.
13. Ibid., 3.
14. Ibid., 4.
15. Ibid., 2.
16. Ibid.
17. Ibid.
18. Ibid., 1.
19. Ibid., 3.
20. Ibid., 6.
21. Ibid., 4.

Environmental Etiquette/Environmental Practice: American Indian Challenges To Mainstream Environmental Ethics[1]

Shagbark Hickory
(in the voice of Johannes Climacus[2])

I CAN THINK OF NO BETTER WAY to acknowledge my very great debt to the writings and person of Paul Shepard than to indicate something of the directions my own thought has taken recently, for they are directly inspired by him. Very few who study hunter-gatherer cultures take them seriously as challenges to Western ways of thought except in very generalized ways (for example, by acknowledging that they did not abuse the land as we have done). Paul is a rare exception. To take him seriously is to take American Indian mindscapes seriously.

At the beginning of my philosophical reflections on environmental ethics stands the figure of sociologist Philip Rieff. Rieff denounced the pervasive moral subjectivism of our time, a version of which he took to be Freud's unfortunate legacy to the Twentieth Century—namely, the view that morality is no more than a set of introjected norms lodged, perniciously, in the superego.[3] In the ensuing Age of Therapy our task became that of overthrowing guilt and coming to understand that the world is a stage, a play of masks. The Age of Therapy, Rieff held, signalled the end of what he called "moral demand systems," conceptions of self and its relationship to the world in which moral demands were understood as objectively real, standing outside the self and demanding obedience. Rieff lamented the decline of moral demand systems—paradigmatically articulated by religious traditions—and saw the Age of Therapy as heralding not a golden age of mental health but a bloodbath, the inevitable result of the Age of Therapy's growing narcissism and lack of moral orientation.

The argument was forceful, and I asked myself whether in fact there were no compelling moral demand systems in the modern secular world.[4] It occurred to me then that environmentalists often spoke as though nature itself was such a moral demand system, that nature itself made moral claims upon us, claims not of our devising but of its devising.[5] The question, then, was how to understand this. As it turned out, there were plenty of ways to understand the claim that we had moral obligations directly to nature without abandoning modern conceptions of ethics.[6] But none of these had the force that Rieff seemed to require of moral demand systems. He seemed to want the other-than-human (for him, God) to speak its demands, and for these commands to be experienced by us as compelling—compelling, not on our terms but on its.[7]

Rieff's insistence can, of course, be understood as simply wrong-headed; modernism can be understood as rightly rejecting moral demand systems not rooted in the human. But there seemed something right about the notion that other-than-human nature might constitute a moral demand system in something like Rieff's sense. This thought can make the various moral worlds of American Indians immediately compelling, for they not only offer moral world views in which other-than humans make moral demands upon humans, but they also offer alternatives both to modern ethical theory and to the sky-god moral demand systems with which we have been all too familiar.

Generally speaking, the natural world is through-and-through a moral world for the American Indian—everything in nature is (at least potentially) deserving of direct moral consideration. And it is a world of compelling moral experience—that is, it is a moral demand system. This moral vision constitutes one of the challenges American Indian thought poses for mainstream environmental ethics. It is worth pointing out that one contemporary environmental ethicist has recently articulated a view of moral considerability that incorporates both aspects of this pervasive American Indian moral view. His work is, I believe, revolutionary for this reason. Central to Cold Mountain Birch's position are the notions that considerability goes "all the way down" and that "deontic experiences" requiring that we behave in certain ways found ethics. Both of these ideas are pervasive in American Indian thought. For Birch, deontic experience does not determine in and of itself what we, morally speaking, ought to do. Deontic experience must be filtered through a culture's fund of considered wisdom. Likewise for the American Indian who must bring back his or her gift of a vision to the tribal elders for interpretation and instruction.[8]

We must nonetheless ask whether the moral worlds of American Indians are acceptable to the modern mind. Universal consideration is a common feature of Indian thought because of the personal nature of the other-than-human. Other-than-human persons can be found everywhere.[9] The personal nature of other-than-humans (coupled with the fact that humans are usually thought to have less power and moral authority than other-than-humans) is also the source of the perception that other-than-humans make binding moral demands upon us.

Can we accept the metaphysics which underlies much American Indian moral thought? Here we have a second challenge to mainstream environmental ethics. And once again the challenge has been taken up. The mathematical biophysicist Evelyn Fox Keller has claimed that

> "Laws" of nature name nature as blind, obedient, and simple; simultane-

> ously, they name [the scientist] as authoritative, generative, resourceful, and complex....By contrast, the conception of nature as orderly, and not merely law bound, allows nature itself to be generative and resourceful—more complex and abundant than we can either describe or prescribe. In this alternative view, nature comes to be seen as an active partner in a more reciprocal relation to an observer....Such a relationship...would require a different style of inquiry, no less rigorous but presupposing the modesty and open attentiveness that allow one to "listen to the material"...[10]

The historian of science Donna Haraway, in related fashion, has claimed that what she calls "situated knowledges" (as opposed to modern conceptions of knowledge as not context dependent) require that we

> grant...the status of agent/actor to the "objects" of the world. Actors come in many and wonderful forms. Accounts of a "real" world do not...depend on a logic of "discovery" but on a power-charged social relation of"conversation." ...Acknowledging the agency of the world in knowledge makes room for some unsettling possibilities, including a sense of the world's independent sense of humor....[W]e are not in charge of the world. We just live here and try to strike up noninnocent conversations.[11]

In a recent interview and follow-up "postscript" Haraway has elaborated on her notion of revisioning the world as Trickster. Here are the relevant passages:

> This potent trickster can show us that historically specific human relations with "nature" must somehow—linguistically, scientifically, ethically, politically, and epistemologically—be imagined as genuinely social and actively relational. And yet, the partners in this lively social relation remain inhomogeneous....We need a concept of agency that opens up possibilities for figuring relationality within social worlds where actors fit oddly, at best, into previous taxa of the human, the natural, or the constructed.[12]
>
> Well, obviously what's going on there is some kind of play with metaphors. In this respect, I'm most influenced by Bruno Latour's actor-network theory, which argues that in a sociological account of science all sorts of things are actors, only some of which are human language-bearing actors....Perhaps only those organized by language are subjects, but agents are more heterogenous. Not all the actors have language. And so that presents a contradiction in terms, because our notions of agency, action, and subjectivity are all about language. So you're faced with the contradictory project of finding the metaphors that allow you to imagine a knowledge situation that does not set up an active/passive split. So you go for metaphors like the coyote, or trickster figure....It's always important to keep the tension of the fiction foregrounded, so that you don't end up making a kind of animist or pantheist claim. How do I then act the bricoleur that we've all learned to be in various ways, without being a colonizer; picking up a trickster figure, for example, out of Native American stories?[13]

It is of more than passing interest that two such highly-respected feminist theorists in the history and philosophy of science—theorists who have thought long and well about objectivity and our (human) epistemic relationship to the other-than-human world in the wake of trenchant criticisms of modernist conceptions of objectivity and epistemology—should converge in their thinking on the notion that other-than-human nature should be considered an active agent (or partner) in the construction of knowledge. I consider this to be one of the most exciting and potentially fruitful ideas in contemporary epistemology.

~ ~ ~

Implicit in the notion of other-than-humans as active agents in the construction of knowledge is another idea, signalling a revolution in our understanding of the relationship between ethics and epistemology. If the object of knowledge is itself an active agent in the construction of knowledge and if all knowledge is "situated" knowledge,[14] then there is not only an ethic (an etiquette) of knowledge acquisition, there is also an ethical dimension to objectivity itself: Objective knowledge is that which (among other things) is co-constructed in accordance with certain ethical practices. That knowledge has an ethical dimension is implicit in American Indian thought and constitutes yet another of the challenges environmental ethicists must face if they are to take American Indian thought seriously.

Many of us who generate theory because it is our job (as academicians, for example) to do so all too often come to think of ourselves as accountable only for the truth of what we say, if that. But this view can be a smoke screen for a profound irresponsibility. Maria Lugones and Elizabeth Spelman, in their critique of the theorizing of Third World people by the dominant culture, point out that theories can not only be characterized as true and false but that they "can be, e.g., useless, arrogant, disrespectful, ignorant, ethnocentric, imperialistic."[15] Very often there is a logic of domination at work in the theorizing of the Other. Reductive theories, for example, bring one order of explanation and meaning under the conceptual control of another order of explanation and meaning, and the objectivist distinction between the knowing subject and the object of inquiry often brings with it both the isolation of the subject from the object and the institution of a hierarchical and exploitative relationship between the two. Evelyn Fox Keller, on the other hand, argues that respect for difference "has a different goal: not prediction per se, but understanding; not the power to manipulate, but empowerment—the kind of power that results from an understanding of the world around us, that simultaneously reflects and affirms our connection to that world."[16] American Indian practices not only emphasize this connection between understanding and empowerment but also suggest the stronger claim that issues of truth and falsity cannot, finally, be separated from such characterizations. And, of course, if other-than-humans are active agents in the construction of knowledge, then all of this applies to our theorizing of nature as well.[17]

~ ~ ~

As I have noted, Donna Haraway suggests that "acknowledging the agency of the world in knowledge makes room for some unsettling possibilities, including a sense of the world's independent sense of humor." This suggestion leads to the next challenge to environmental ethics posed by American Indian thought.

In *The Comedy of Survival* (with an introduction by Paul Shepard) Joseph Meeker contrasts the tragic mode with the comic mode in Western literature, arguing that "comedy grows from the biological circumstances of life." "[S]piritual and artistic creativity are not special powers provided so that humans can transcend the natural world, but features of human biological development useful for connecting humanity more deeply with the world."

> [S]tructures in nature...reveal organizational principles and processes which closely resemble the patterns found in comedy. Productive and stable ecosystems are those which minimize destructive aggression, encourage maximum diversity, and seek to establish equilibrium among their participants—which is essentially what happens in literary comedy. Biological evolution itself shows all the flexibility of comic drama and little of the monolithic passion peculiar to traged.... Like comedy, mature ecosystems are cosmopolitan. Whatever life forms may exist seem to have an equal right to existence, and no individual needs, prejudices, or passions give sufficient cause to threaten the welfare of the ecosystem structure as a whole. Necessity, of course, is real. All must eat and in turn be eaten, storms must come and go, and injustices must occur when so many rightful claimants contend. But that is just the point: comedy and ecology are systems designed to accommodate necessity and to encourage acceptance of it, while tragedy is concerned with avoiding or transcending the necessity in order to accomplish the impossible.[18]

The challenge posed by many American Indian cultures is in their exemplification of cultures organized in the comic rather than in the tragic mode.[19] American Indian trickster figures and Pueblo clowns, for example, are central figures in rich traditions in which human relations with nature are social and actively relational, in which people do not think of themselves as "in charge of the world," in which they simply "live...and try to strike up noninnocent conversations," as Donna Haraway puts it.

The most salient (and at first sight paradoxical) feature of Pueblo clowns and the various trickster figures in American Indian oral traditions is their status as, on the one hand, socially peripheral and marginal and, on the other, central and predominant.[20] Clowns and tricksters bear some resemblance to the Fool in Western literature (e.g., in Shakespeare), who "usually stands apart from the main action of the play, having a tendency not to focus but to dissolve events.... The serious hero focusses events, forces issues, and causes catastrophes; but the Fool by his mere presence dissolves events, evades issues, and throws doubt on the finality of fact."[21] Clowns and trickster figures, however, play a central role in many American Indian cultures. Babcock-Abrahams reads the trickster as a deconstructive figure at the heart of culture. Trickster tales, she argues, "express the generative situation of ambivalence and contradictions that the very basis of culture engenders. Seeming undifferentiation and ambivalence are characteristic of mediating figures, and it may well be that the mediating figure of Trickster…is created in response to a present and constant perception of opposition, of difference essential to human constructs…"[22] As Paul Radin, Babcock-Abrahams and others point out, trickster tales satirize the most sacred and cherished of tribal institutions. This is so for Pueblo clowns as well.

As deconstructionists, clowns and tricksters operate in the comic rather than in the tragic mode. To distinguish deconstruction as it figures in American Indian culture from various forms of postmodernist deconstruction, we might coin the term "deconstructive burlesque": deconstruction accomplished in the comic mode with the intent of achieving that form of cultural affirmation appropriate to comedy:

> "Marginality" may also be distinguished into its comic and tragic modalities…. In contrast to the scapegoat or tragic victim, trickster belongs to the comic modality or marginality where violation is generally the precondition for laughter and communitas, and there tends to be an incorporation of the outsider, a levelling of hierarchy, a reversal of statuses.[23]

Further, Babcock-Abrahams explains, trickster tales "are preoccupied with those areas between categories, between what is animal and what is human, what is natural and what is cultural. Trickster and his tales exemplify this preoccupation, for at the center of his antinomian existence is the power derived from his ability to live interstitially, to confuse and to escape the structures of society and the order of cultural things." More importantly, trickster expresses the ambiguous and paradoxical nature of power so derived. While trickster's power endows his group with vitality and other boons, it also carries the threat and the possibility of chaos. His beneficence, though central, results from the breaking of rules and the violating of taboos."[24] Sam Gill points out that the defiling acts of clowning

> …present the experience of chaos, of no order at all, of nondifferentiation. The great [deconstructive] danger risked in these activities is accompanied by the [constructive, generative] acquisition of power—for in transgressing such ominous boundaries, the clowns engender the very forces of differentiation that distinguish order from chaos and thereby create a meaningful world…. Defiling clowns are often considered more powerful than the shamans.[25]

For the American Indian, language—understood in a broad sense so as to include ritual and ceremony in addition to storytelling and myth—is world generating and world renewing. As Sam Gill puts it, "Native Americans do not restrict language to its capacity to describe the world: they recognize that, from one perspective, it is the world."[26] The world—the world created and sustained by language—is therefore fragile. (It is also plural. American Indians do not seem to have a notion of The Truth. When a culture dies, a world dies, though other worlds remain.) Great cultural means must be taken to call it into existence and maintain it in existence. This is the great calling of humankind, and other-than-human-persons decide to provide for us, to help us, for they like to hear the stories we tell of them.

In addition to this great calling—or as a corollary of it—is the religious necessity of experiencing the world's fragility, of understanding it as arbitrary. Arbitrary but necessary: the world a culture constructs is the world in which it lives; it is the world in which the members of the culture have their existence and identity, their very being. This explains in large measure the central importance of the trickster figures and clowns. As culturally-central figures, tricksters and clowns serve more than a deconstructive function. In addition to "dissolving events," "throwing doubt on the finality of fact," and revealing the socially-constructed nature of myth, ritual, and social structure tricksters, clowns (and other tribal institutions as well) serve to deepen tribal understandings of these targets of satire.

Accounts of clowns in the Southwest are very like descriptions of Zen masters in the Buddhist tradition.[27] The work of the Zen master is at one and the same time deconstructive and a point of entry for satori, enlightenment. The religious deepening called satori is actually accomplished by the deconstruction of received religious doctrine. It is a move beyond doctrine to the world itself—with the paradoxical understanding that the world is a world "made of words"[28] (in the broad sense). These phases of religious deepening are captured nicely in the idea that before undergoing religious training in Zen a mountain is just a mountain; while coming to understand Buddhist doctrine a mountain is no longer a mountain; with enlightenment a mountain is once again a mountain. This world and Nirvana are one and the same.

What all this suggests is not only a sophisticated parallel to the deconstructive techniques of some contemporary philosophers and literary critics, but also an equally sophisticated understanding of myth, ritual, and ceremony. Herein lies yet another challenge to mainstream environmental ethics. Clowns and tricksters work the boundaries of the constructed world of Indian existence, stressing both its constructed nature and its reality.

Deconstruction has enabled American Indians to avoid two related pitfalls of modernism: literalism and foundationalism.[29] American Indian cultures have managed to retain what might be paradoxically described as a"postmodern"relation to tribal myths, rites, and social institutions without having gone though a modernist phase—a condition which makes for a "postmodernism" which lacks the troubled and self-conscious nature of much Western postmodernism.

As an example I cite Sam Gill's discussion of the Hopi Kachina cult initiation. At the time of initiation Hopi children for the first time discover what had been a carefully guarded secret—namely, that the kachinas they have often seen in their village are actually their own relatives. The result, Gill says, is often severe disenchantment, leaving the children "at the very threshold of a formal religious life, with serious doubts about the reality of the spiritual figures whom they had believed were essential to the Hopi way of life." Deconstruction, however, is conjoined with the simultaneous deepening of religious and social sensibility and understanding. "The experience of disenchantment strikes a deathblow to the naivete that is characteristic of the uninitiated. . . . This knowledge establishes an agenda of religious inquiry and a keen interest in pursuing it. The meaning of one's life depends upon it."[30]

American Indian cultures, so far as I know, make no claim to privileged status (or truth) with respect to other cultural constructions. Pluralism in these cultures is not a matter of "toleration," it is an absence of literalism and foundationalism. American Indian "postmodernism" is not modernist in any sense. This is precisely what makes it so difficult for us to understand the sense in which they "believe" their myths.

A woman in my American Indian Philosophies class brought in an Ojibwa dreamcatcher and related the following story about her eight year old son. He had recurring and very frightening nightmares. Consultation with various physicians, psychologists, and psychiatrists were of no avail. In school the boy heard about Ojibwa dreamcatchers, which, when placed above a person's head when asleep catch bad dreams and hold them to be burned off by the morning sun. He insisted that his parents get him a real dreamcatcher, one made by an Ojibwa. They did this. The nightmares ceased.

The important thing to note here is that the dreamcatcher worked; whatever else might be said about the case must be faithful to that fact. To reduce it to a psychological account seems to elide that fact in some important way. To insist, for example, that we talk about the dreamcatcher in the language of psychology and that his parents relate to their child concerning the nightmares only in the language of psychology would be disabling. In fact, that was tried and did not work. In response, it might be said that, even so, it is the psychological account which is the true one and that we can account for the fact that when the dreamcatcher was placed above the boy's head the nightmares ceased in psychological terms. But this response is foundationalist, an insistence upon a particular vocabulary as privileged since it corresponds with reality, the way things really are.[31]

Suppose one were to insist that a computer just is a complex device for shuttling electricity and activating printers in response to taps on the keyboard and that, therefore, only the language of the electrical engineer is permissible in talking about what the computer does—no talk of computers adding and subtracting, no talk of drafting an essay on the computer is permissible. This would be severely disabling; we could not use computers as we in fact do. If we have no problem with multiple, distinct vocabularies with which to talk about computers we should, in principle, have no problem with dreamcatcher talk. None of this, of course, precludes the possibility that Ojibwas might come to have good reasons for dropping dreamcatcher talk. My only point is that foundationalist reasons aren't good reasons. In the dreamcatcher case we would be foregoing something of considerable importance by dropping the dreamcatcher vocabulary, just as the Hopi children in Sam Gill's example of Hopi disenchantment in the Kachina cult initiation, sketched above, would if, after the experience of seeing that the kachinas are actually their relatives in masks, they insisted that there are no kachinas. "My relatives are kachinas" reaches deeper than "There are no kachinas." So "The dreamcatcher catches dreams" reached deeper than the available psychological accounts.

In a discussion of Santa Claus with my American Indian Philosophies class, a class with many returning adult students with children, it was noted that children respond in basically two (quite different) ways to their disenchantment with respect to a literal understanding of the Santa Claus myth. One group responded: "There is no Santa Claus." The other responded: "I [the parent or parents] are Santa Claus." The discussion revealed the latter response to be much more sophisticated than the former (literalist) response. It opened up not only a much richer understanding of the parent(s), the child, and the relationship between them, but it opened the child to a much richer sense of language, story, myth, and ceremony. Several wonderful stories were told concerning children who responded in the second (Hopi) way to disenchantment. John Farella has noted that

> Natives have been assumed to be fundamentalists. Along with this has gone the assumption that they are cognitively operating at a very literal level.... What has happened is that the basic premise underlying our research on the religion of others—that we must "respect" their beliefs—has produced unquestioned literal interpretations that can be treated only as if they were beliefs, for from our perspective, they cannot be fact."[32]

Genuine respect for native philosophies engage them on their own terms rather than treating them as artifacts to be understood in literalist terms and preserved and "respected" as such. Literalism has its roots in foundationalism.[33] Since American Indians—those not acculturated to Western epistemology, that is—are not foundationalist, they are not fundamentalists either.[34]

It isn't that there are no Indian literalists. There are. We have created them. Farella argues that literalism has emerged within American Indian cultures as a result of the loss of the myths of origins for particular rituals. With the loss of myths of origin comes a loss of the philosophical understanding embedded in the rituals. The rituals are no longer transparent vehicles of power and thought. With this loss of philosophical understanding comes fear. The only thing left is ritual. A general sense of the ritual may be there, but the details are no longer transparent vehicles of power and understanding. Dependence upon the details of the ritual-getting it right—becomes paramount, for no one knows what may happen when details are omitted or altered. As Vine Deloria put it, "Rituals lose their efficacy because they are performed within a secular time. . . . They soon take on the aspect of mechanical adjustments made to solve problems which occur within that kind of time."[35]

This is parallel to dependence upon the word in contemporary literalism. There is reason to believe that fetishism of the word constitutes the first appearance of literalism and that contemporary American Indian literalist attitudes with respect to ritual are derivative, the result of the destruction of the link between myth and ritual within indigenous cultures by literate cultures. Prior to the rise of literate culture the word (in the broad sense) was a medium, transparent to power and experience. They created the world—they were powers in the world. One lived in the ambience of myth and ritual as one lived in the ambience of the world itself.

(To get a taste of what it might be to live in nonliteralist fashion within the ambience of myth, read Gary Snyder's "The Woman Who Married a Bear."[36] After relating the tale as told by Maria Johns, Snyder provides a "commentary" which in fact amounts to a retelling of the story in his own voice but within the ambience of the myth; that is, he tells the story as Maria Johns told it, not pretending to be her; he tells it as one who has led his own particular life (and therefore with his particular idiographic twist) and yet as Maria Johns presumably told it—within the ambience of the myth, not as an artifact to be appreciated in a certain way or understood from the point of view of a privileged (truth-conveying) discourse. The difference between Snyder's telling and Maria John's is the difference between two distinct idiographic routes through the world, not the distinction between the telling of the story by a "believer" and the appropriation or colonizing of that story into, or by, a privileged discourse. One senses, in savoring Snyder's telling of the story, just how easy it might be to become again a mythic people—and also how hard it would be: the resistance wells up, the desire, no matter how sympathetic one might be, to tell the story "objectively," to relate it as a story told by other and wiser—because innocent—people and to extract lessons from it, lessons told in the Language of Truth.)

Western literalism—by which I intend to include the foundationalist epistemologies that have characterized the modern period—has been shaped by the subject-object split, locating the world-as-object outside consciousness and both representations of the world and valuation within the human subject. These epistemologies

> turn everything into a resource for appropriation, in which an object of knowledge is finally itself only matter for the seminal power, the act, of the knower.... Any status as agent in the productions of knowledge must be denied the object. It—the world—must, in short, be objectified as a thing, not as an agent; it must be matter for the self-formation of the only social being in the productions of knowledge, the human knower.[37]

With the failure of epistemologies based on this radical division to regain the world—as, for example, with the correspondence theory of truth—social constructionism gained the day. But the notion of the social construction of knowledge, holding, as it most often does, that only human subjects are agents in the construction of knowledge, leaves the modern subject-object split firmly in place. Social constructionism all too often leaves modern objectivism intact since it so often leaves the modern subject in place as the constructor of knowledge.

Haraway's suggestion (and the pervasive American Indian view) that we understand the world as actor and agent in the construction of knowledge is a suggestion that we break finally and completely with modern epistemologies. The following, from a study of a contemporary Chipewyan group by the anthropologist

Henry Sharp, graphically illustrates the contrast between Western and Indian epistemologies I have been developing:

> July 24 brought an unwelcome surprise in the form of a [Royal Canadian Mounted Police] visit.... We are now conditioned to accept that the symbols, ideas, and language of alien cultures are ways of knowing the environment within which they dwell, but we have conveniently managed to subordinate the significance of that understanding to our quest for objectivity. These things are not passive ways of perceiving a determined positivist reality but a mode of interaction shared between the dene [Chipewyan] and their environment. All animate life interacts and, to a greater or lesser degree, affects the life and behavior of all other animate forms. In their deliberate and splendid isolation, the Chipewyan interact with all life in accordance with their understanding, and the animate universe responds. White Canada does not come silently and openly into the bush in search of understanding or communion; it sojourns briefly in the full glory of its colonial power to exploit and regulate all animate being and, foremost of all, the dene. It comes asserting a clashing causal certainty in the fundamentalist exercise of the power of its belief. It talks too loudly, its posture is wrong, its movement harsh and graceless; it does not know what to see and it hears nothing. Its presence brings a stunning confusion heard deafeningly in a growing circle of silence created by a confused and disordered animate universe.[38]

When we see American Indian myth, stories, ceremonies, and society without the distorting lens of Western literalism we see that they encode a wealth of social, psychological, and ecological information in a framework that truly understands the world as actor and agent in the genesis of value and knowledge.[39] In an earlier paper I described American Indian thought as carried in language that grows out of experience and articulates it, language intermediate between self and world, their intersection, carrying knowledge of both, knowledge charged with valuation and instruction.[40] Paul Shepard suggests this is language in which "the clues to the meaning of life [are] embodied in natural things, where everyday life [is] inextricable from spiritual significance and encounter."[41] The vision received in the American Indian vision quest, for example, is a culturally mediated intersection of self and world. It is a gift and must be located in the world. It is important not to conceive of these images as projections; they are intersections, encounters. This way of putting it acknowledges both our active construction of reality and nature's role in these negotiations. We should say with Aiken that "The landscape and the language are the same, /And we ourselves are language and are land."[42]

To understand the world as Coyote, as Trickster, as "witty agent and actor," is to understand that the world eludes sense as a joke does, that no system (whether ethical, metaphysical or epistemological) captures it, contains it.[43] The trickster wanders on the margins, outside society's structures, though engendering society and remaining as deconstructive ambiguity and questioning within it.

American Indian thoughtways offer a strong challenge to our culture's attitudes and practices concerning the food we eat.

In the West, either no moral consideration is deemed necessary in relation to the plants and animals we eat or, if moral consideration is given, our response is most often radically dichotomous. Consider the common inference from the claim that the domination and exploitation of animals is wrong to the conclusion that vegetarianism is morally required.[44] The domination and exploitation of plants (not to mention abiotic entities and ecosystems as such) is apparently unproblematical for many who draw this inference.[45] The world is divided into those things which enjoy moral standing (animals) and those which do not. Does this make good moral sense? I do not think that the insistence on vegetarianism by itself gets at the critical moral issues posed by the modern practice of raising and slaughtering animals for food.

Consider plants. It seems clear that the domination of nature includes the domination of plants as well as the domination of animals. The treatment of plants within modern industrial agriculture seems to be an instance of the domination of nature quite on a par with the modern factory farming of animals. Should we not oppose both? If so, what does this imply for our eating practices?

As ecofeminists such as Karen Warren argue,[46] it is the interconnection between the various modes of domination that is the crux of the problem. It is the logic of domination that is the problem. So long as we leave domination in one place intact it infects efforts at liberation in other places. If we simply do not see the liberation of plants as an issue, then in one way or another we are going to leave the logic of domination in place in our relationship to animals (as well as in our relationship to plants, ecosystems, etc.) even if we become vegetarians. Vegetarianism is often not just incomplete but also problematic, for reasons stated above—namely, when it is framed in the moral categories used to distinguish a morally considerable "we" from a nonconsiderable "they."

If the "No exploitation!" rule implies vegetarianism when applied to animals, it would seem to imply "No eating!" if it is also applied to plants. The way out envisioned by many (most?) contemporary theorists is to dichotomize existence into the morally considerable on the one hand and "mere things" that do not belong to the moral club on the other. If we are not to dichotomize existence in this way, and if we see the absurdity of the "No eating!" conclusion, we must rethink what "No exploitation!" implies in the case of other-than-humans.

Here we run up against an apparent paradox of existence: we have moral obligations to the very food we kill and eat, whether animal or vegetable. American Indian thought embraces this paradox and centralizes it. American Indian thought suggests that correct moral relationship to animals requires that, as Cold Mountain Birch has put it, we take moral considerability "all the way down"—e.g., to plants and rocks—and that, as an extension of what Donna

Haraway suggests concerning the objects of knowledge, we consider the objects of moral relationships (including rocks) as actors and agents in the construction and understanding of these relationships.

For most or all American Indians food (plant as well as animal) is kin. Relationships to plants and animals as, on the one hand, food and, on the other hand, kin creates a tension[47] which is dealt with mythically, ritually, and ceremonially, but which is never denied. It is this refusal to deny the dilemma in which we are implicated in this life, a refusal to take the way of bad faith, moral supremacy, or self-deception which constitutes a radical challenge to our relationship to our food. The American Indian view that considerability goes "all the way down" requires a response considerably more sophisticated than those we have seen in the West, which consist either in drawing lines of moral considerability in order to create an out-group, or in constructing hierarchies of considerability creating de facto out-groups in particular cases.

~ ~ ~

Paul Shepard is suspicious (to say the least) of agriculture. Though he is charitable towards older horticultural practices, he is nonetheless suspicious on the grounds that once humans started down the agricultural road we eventually found ourselves in the midst of neolithic horror stories, both in the New World and the Old. Is the appropriate response to this to sit at home eating feedlot steaks and monoculture wheat while eschewing the agricultural vision of a Wendell Berry or a Wes Jackson? Or do these visions appear in a different light when we look forward to the neolithic from earlier hunter-gatherer and horticultural practices (as does Shepard) than they do when we look back through the neolithic to the paleolithic. Looking from the latter perspective we might perhaps be able to construct a mode of agriculture which is truly an "enabling practice."[48]

Where we need to look for some challenge to Western practices, I suspect, is to smallscale horticultural cultures such as those in the American Indian Southwest. It is at least possible that here we can find clues as to how a culture can organize itself in relationship to its food in ways that do not involve domination although they do involve the domestication of plants and sometimes animals as well. Domestication is a given in the immediate future and it therefore behooves us to look closely at those examples of indigenous cultures that seem to have evolved relations to their food (within domestication) that do not involve notions of domination but, rather, notions of reciprocal kinship.

The Hopi, for example—with their notion of cooperation without submission, their understanding that humans emerged into the world as corn does, their treatment of corn as new born persons, their view that music must accompany the growth of the corn, that the ancestors return as life-giving rain, and that to eat corn is to plant history and song into the person—do not seem to have forged a culture structured by domination, either to corn or to one another. Such a nurturing relationship to domesticated plants and animals, in which they are understood as kin and in which participation and reciprocity, rather than domination, characterizes the relationship, does not seem inherently problematic: it is simply a matter of carrying what is good, what is nurturing, in our (human) domestic relationships to one another over into our relationship to domesticated other-than-human persons. If nondomination is a real possibility within human domesticity, then it is a real possibility in our domestic relations with other-than-human persons.

This is not to say that a thoroughly domesticated world is desirable or even possible in any way that is consistent with human health and well-being. Indeed, it is one of Paul Shepard's enduring lessons that (to once again borrow terms from Gary Snyder) our good and the world's can come to be (again) only within the wider context of an etiquette and practice of the wild.

Notes

[Editor's note: Shagbark Hickory is a pseudonym for Jim Cheney.]

1. The title of this essay acknowledges my debt to another of our elders and keepers of the lore: Gary Snyder. I would like to acknowledge the help and encouragement of Cold Mountain Birch, Daddy Longlegs, and all the other-than-human persons who attended (or at least tolerated) the Zen Lite Prairie, Mountain, and Great Lakes Conferences of 1991-93. It was without doubt Inyan (Stone) who made his presence felt so strongly at Zen Lite II and who continues to watch beside our trails. The power of Sleeping Bear, who presided over Zen Lite III, surges deep in the Dreamtime. Aurora Borealis blessed Zen Lite I. We humans hope that the other-than-humans enjoyed the stories. Max Oelschlaeger's comments on an earlier draft were invaluable.
2. Johannes Climacus is the voice Kierkegaard used in his *Concluding Unscientific Postscript to the Philosophical Fragments,* where he portrayed Christianity favorably in relationship to academic philosophy, but did so in the language of philosophy rather than in the language of Christianity. The result was intentionally paradoxical and ironic: Though portrayed in the language of philosophy, Christianity was portrayed as superior to and as transcending in a fundamental way the categories inherent in academic philosophy. The result highlighted the incommensurable differences (to Kierkegaard's mind) between the prevailing philosophical discourse and Christianity. A similar problematic, paradoxical, and ironic tone pervades the present essay. This should be borne in mind throughout. Max Oelschlaeger has reminded me (see"Earth-talk: Conservation and the Ecology of Language," a paper presented at the Fifth World Wilderness Congress, Tromsø, Norway, September, 1993) of Walter J. Ong's observation that "Writing is indeed the seedbed of irony, and the longer the writing (and print) tradition endures, the heavier the ironic growth becomes" (Ong,

Orality and Literacy: The Technologizing of the Word [London and New York: Routledge, 1988]). It is particularly important in this postmodern era that this irony become self-conscious and that we become adept at listening as ironists to texts which can fruitfully be understood only as irony. Many postmodern texts are read as implicitly modernist because they are not read with this irony in mind. I submit that any postmodern text can plausibly be read as implicitly modernist, or as having a modernist subtext—if it is not read as ironical.

3. Philip Rieff, *Fellow Teachers* (New York: Dell Publishing Co., 1972).
4. Various fundamentalist religions did, of course, offer moral demand systems, but often enough these brought bloodbaths of their own in their wake.
5. It makes no difference at this point whether we understand the objective reality of nature-as-a-moral-demand-system as objective in a modernist sense or whether we, in a postmodernist vein, understand it as a social construction.
6. I am thinking of writers such as Tom Regan, Peter Singer, Paul Taylor, and Kenneth Goodpaster.
7. Kant's categorical imperative is compelling, certainly, but on our terms. That is, the imperative is rationally compelling, and rationality, for Kant, defines us transcendentally.
8. Cold Mountain Birch, "Moral Considerability and Universal Consideration," *Environmental Ethics* 15 (Winter 1993):313-32. [Editor's note: "Cold Mountain Birch" is a pseudonym for Thomas H. Birch.]
9. The anthropologist, A. I. Hallowell, notes that for the Ojibwa the class of persons is far broader than the class of humans. Hallowell insists upon the empirical and nondogmatic basis for this notion. He relates the following story: "Since stones are grammatically animate, I once asked an old man: 'Are all the stones we see about us here alive?' He reflected a long while and then replied, 'No! But some are.' This qualified answer made a lasting impression on me. And it is thoroughly consistent with other data that indicate that the Ojibwa are not animists in the sense that they dogmatically attribute living souls to inanimate objects such as stones. The hypothesis which suggests itself to me is that the allocation of stones to an animate grammatical category is part of a culturally constituted cognitive 'set.' It does not involve a consciously formulated theory about the nature of stones. It leaves a door open that our orientation on dogmatic grounds keeps shut tight. . . . The crucial test is experience." In this example Hallowell is discussing the "animate" category, not that of "persons." Similar considerations apply in the case of persons, however, as Hallowell's work clearly testifies.

See A. I. Hallowell, "Ojibwa Ontology, Behavior, and World View," in Dennis and Barbara Tedlock, eds., *Teachings from the American Earth: Indian Religion and Philosophy* (New York: Liveright), 147-8. See also his "Some Empirical Aspects of Northern Saulteaux Religion," *American Anthropologist* 36 (1934):389-404, and Mary B. Black, "Ojibwa Taxonomy and Percept Ambiguity," *Ethos* 5 (Spring 1977):90-118.

10. Evelyn Fox Keller, *Reflections on Gender and Science* (New Haven and London: Yale University Press, 1985), 134.
11. Donna Haraway, "Situated Knowledges: The Science Question in Feminism and the Privilege of Partial Perspective," *Feminist Studies* 14 (Fall 1988):596, 593-4.
12. Donna Haraway, "The Actors Are Cyborg, Nature Is Coyote, and the Geography is Elsewhere: Postscript to 'Cyborgs at Large,'" in Constance Penley and Andrew Ross, eds., *Technoculture* (Minneapolis: University of Minnesota Press, 1991), 21.
13. Donna Haraway, "Cyborgs at Large: Interview with Donna Haraway," in Penley and Ross, eds., *Technoculture,* 3.
14. There is another challenge taken up by Haraway's notion of "situated knowledges," but since it has been taken up in many quarters already it merits only passing mention here. This is the challenge to modern conceptions of knowledge in the idea that knowledge is contextual.
15. Maria C. Lugones and Elizabeth V. Spelman, "Have We Got a Theory for You! Feminist Theory, Cultural Imperialism and the Demand for 'The Woman's Voice'," in Marilyn Pearsall, ed., *Women and Values: Readings in Recent Feminist Philosophy* (Belmont, Cal.: Wadsworth Publishing Company, 1986), 26.
16. Keller, *Reflections,* 166.
17. This is a particularly ironic passage, written, as it is, by the sympathetic but academic Johannes Climacus. It implies the bad faith of its own earnest account.
18. Joseph W. Meeker, *The Comedy of Survival: In Search of an Environmental Ethic* (Los Angeles: Guild of Tutors Press, 1980), 38, 155, 41, 43-4. Meeker's ecological theory is somewhat dated, but his argument still holds—is perhaps strengthened—when the ecological theory is updated.
19. Tragedy portrays "the enormous human capacity for creating and enduring pain, for following a passion to its ultimate end, for employing the power of mind and spirit to rise above the contradictions of matter and circumstance even though one is destroyed by them." Ibid., 37-8.
20. See Barbara Babcock-Abrahams, "'A Tolerated Margin of Mess': The Trickster

and His Tales Reconsidered," *Journal of the Folklore Institute* 11 (1975):155.

21. Enid Welsford, as quoted in ibid., 154.
22. Ibid., 163-4.
23. Ibid., 153.
24. Ibid., 147-8.
25. Sam Gill, *Native American Religions* (Belmont, Cal.: Wadsworth, 1982), 76.
26. Sam Gill, "The Trees Stood Deep Rooted," in D. M. Dooling and Paul Jordan-Smith, eds., *I Become Part of It: Sacred Dimensions in Native American Life* (San Francisco: HarperCollins, 1992), 23.
27. See the accounts by Barbara Tedlock ("Boundaries of Belief," 124-38), Emory Sekaquaptewa ("One More Smile for a Hopi Clown," 150-7), and Sam Gill ("Disenchantment," 106-19) in Dooling and Jordan-Smith, eds., *I Become Part of It.*
28. Max Oelschlaeger suggests that Dogen might appropriately interject "What world? What words?"
29. Literalism is the view that there is a privileged understanding of words such that this understanding constitutes their literal meaning. Truths about the world are expressed in literal language. If a sentence is not literally true, then it is not true, period. Literalism is fairly obviously linked to foundationalism. But if language is metaphor through and through, then there are no foundations and all knowledge is metaphorical (as well as socially constructed).
30. Gill, *Native American Religions*, 92.
31. Inevitably, the charge will be made that the claim that the dreamcatcher worked is an instance of post hoc reasoning. This charge misses the point of the example. No claim is made about causal connections—e.g., that the dreamcatcher as a physical object causes the destruction of bad dreams. The point is that in the "world" constructed in part by dreamcatcher language the dreams no longer occurred, as they had prior to its construction. See n. 35, below, and the passage to which it is appended.
32. John R. Farella, *The Main Stalk: A Synthesis of Navajo Philosophy* (Tucson: The University of Arizona Press, 1984), 8-9. See also 4-9 and 51-3.
33. See n. 29, above.
34. Farella notes that Navajo healers have no trouble accepting both the native and Western understandings of illness without reduction of one to the terms of the other, whereas Western physicians find this unintelligible. It is the Western physician who is the fundamentalist.
35. Vine Deloria, Jr., "Out of Chaos," in Dooling and Jordan-Smith, eds., *I Become Part of It*, 265. Deloria's point has obvious relevance to the post hoc

charge discussed in n. 31, above.

36. Gary Snyder, "The Woman Who Married a Bear," in *The Practice of the Wild* (Berkeley: North Point Press, 1990), 155-74.
37. Haraway, "Situated Knowledges," 592.
38. Henry S. Sharp, *The Transformation of Bigfoot: Maleness, Power, and Belief Among the Chipewyan* (Washington D. C.: Smithsonian Institution Press, 1988), 144-5. Sharp imports something of his own Western animate-inanimate distinction into his discussion.
39. I would point out the visual metaphors at work in Johannes Climacus's statement.
40. Jim Cheney, "Postmodern Environmental Ethics: Ethics as Bioregional Narrative," *Environmental Ethics* 11 (Summer 1989):117-34.
41. Paul Shepard, *Nature and Madness* (San Francisco: Sierra Club Books, 1982), 6.
42. The lines from Conrad Aiken are quoted in Edith Cobb, *The Ecology of Imagination in Childhood* (New York: Columbia University Press, 1977), 67.
43. The thoughts in this sentence were suggested by Daddy Longlegs at the 1992 Zen Lite Mountain Conference. [Editor's note: "Daddy Longlegs" is a pseudonym for Anthony Weston.]
44. For simplicity of exposition, I ignore the fact that many theorists place only some animals in the privileged group.
45. Many, but by no means all. Many who place special emphasis on vegetarianism would not disagree with the analysis I set forth here. Vegetarianism, for these other theorists and activists, is placed within the broader context of considerability "all the way down." Marti Kheel, I think, is an example of a vegetarian who sets her views and practice within this broader context. Peter Singer is an example of a vegetarian who arrives at his views and practice by a process of dichotomization.
46. See Karen J. Warren, "The Power and the Promise of Ecological Feminism," *Environmental Ethics* 12 (Summer 1990):125-46.
47. For a dissenting opinion, see Calvin Luther Martin, *In the Spirit of the Earth: Rethinking History and Time* (Baltimore: Johns Hopkins University Press, 1992), 84-87, who claims that the transformation of moose into human, human into moose in the hunt prevents a moral dilemma from arising, though "I must hunt walrus with courtesy, with restraint, and with the animal's permission, because I am walrus."
48. This term is from Daddy Longlegs' "Before Environmental Ethics," *Environmental Ethics* 14 (Winter 1992):321-38. This article, like that of Cold Mountain Birch cited in n. 8, above, is truly revolutionary, a genuine challenge to mainstream environmental ethics.

Paul Shepard's "The Wild" and Biocentric Holism: A Convergence of Ends

Laura Westra

INTRODUCTION

> At its unnoticed level what is happening is that the genetic prestructure of the optimal or "normal" human individual is being recognized.
>
> —*Paul Shepard*

"WILDERNESS IS WHERE MY GENOME LIVES" was the title of Paul Shepard's spellbinding presentation at the 5th World Wilderness Congress (held in Tromsø, Norway, September 1993). Speaking of the decoding of the human genome, which he characterizes as sustaining our "technological hubris," he pointed out the implicit but unrecognized point cited above. It is a thoroughly Aristotelian position to acknowledge that there is indeed, at least potentially, a "normal" or "optimal human individual." Modern and post-modern thinkers have emphasized in contrast that what we are intrinsically is "freedom," understood as unlimited choices. Both the Aristotelian and the Thomist would agree with Shepard today in terming this belief "the bizarre, modern social hallucination of 'no limits.'"[1] The "no limits" stance is also a deeply "unecological cultural stance," and it is from that standpoint that I have attacked it in my defense of integrity and of the principle of integrity. Shepard's anthropocentrism cannot be described as "weak," like that of Bryan Norton, for instance. His anthropocentrism is strong and uncompromising, but it is also entirely reconcilable with a biocentric (non-anthropocentric) position,[2] because it is basic, foundational, at a level below or beyond that of culture-bound conceptions of the term.

In my discussion of Shepard's position in the "wilderness" paper and elsewhere, I intend to render explicit both the similarities between biocentric holism (based on Aristotle and complex systems theory) and Shepard's position, and some of the differences, particularly on the question of "domestication" of plants and animals (hence of appropriate human diet). On the whole, I am delighted to be able to affirm that the similarities far outnumber the differences, as I will show below.

"WILDERNESS IS WHERE MY GENOME LIVES"

What is Shepard's position in regard to the wild? He says, "If we follow the definition of "domestic" as a population modified by the deliberate control of breeding by humans, then we ourselves are genetically wild."[3] This means far more than admitting we are descended from apes and other wild creatures through evolution. It means that what we are, intrinsically, is very much a part of that "biological wilderness" or biodiversity we are committed to saving. We are no more originally and intrinsically "domesticated" than cows and pigs (or were, as buffalo or wild boars) are, and in fact, "modern culture conceals this but does not alter it."[4] It is the greatest delusion to believe we are apart, different in kind and separate from the rest of nature, hence that all manner of "unnatural" modes of existence will suit us just as well, and probably better than an "animal" existence, governed by instinct, evoking "images of slavering brutes, as an alternative to Rousseau's fictions of innocence and integrity."[5]

The perennial presence of the "Pleistocene" is the benchmark against which we must test the viability and appropriateness of (a) our food choices; (b) our social institutions; (c) our cultural goals; and (d) our very lifestyle. Examples of (a) may be the re-acceptance of death through "participating once again in the great round," whereas in "agriculture participation turns into manipulation"; this also means recognizing the importance of hunting and meat-eating to the diet of "wild man." For (b), we must recognize that huge centralized institutions do not represent "progress" over the way a few people (a maximum of two times twelve, or twenty-five) interact in basic tribal government.[6] Items (c) and (d) above are illustrated by our present unecological forms of living and cultural goals, manifest in lifestyles marked by a lack of humility for other than human life-forms.[7]

The pressures engendered by the latter, Shepard prophetically suggests, will be precisely the point at which health concerns (hence prudence) will force us to recognize our wild origins once again, and turn our back to unsustainable lifestyles. "Mass society" and the pursuit of "virtual but never real reality" are what needs to be transcended in the name of survival, the survival of humans as individuals and as a species.[8] Our salvation lies in the fact that we need not "return to the cave," a hard to defend position. On the contrary, Shepard says, "A culture is an assemblage. It is not a monolith."[9] There are many critical aspects of the "Pleistocene" which must and can be recognized, brought out and re-accepted, without thereby returning to a lifestyle long past. (Shepard provides a list that can be found at the end of his Tromsø paper.)

This is a strong indictment of our blindness and misdirection, and papers and theses can be written to address each and every one of his "accessible components of a Pleistocene paradigm." I will limit myself to a far more modest task, that of reconciling some major "components" he suggests to an ethic of biocentric holism, and rejecting some others, particularly in the areas of "domestication" and human diet, from the same holistic perspective.

THE FOUNDATIONAL VALUE OF WILDERNESS FROM THE PERSPECTIVE OF THE PRINCIPLE OF INTEGRITY

I have argued that ecosystem integrity can be understood as co-extensive with "the wild," as supported for instance by Reed Noss' work in the Wildlands Project. I have also argued that not only the goals of conservation biology (that is, the support of biodiversity and diverse habitats in large core areas, and in appropriately scaled buffers and corridors), but also the goals of sustainable agriculture, the defense of human rights to life, and freedom from environmentally induced morbidity support the absolute requirements of integrity. I have defined ecosystem integrity as a goal compatible with ecosystem health, but separate from it.[10]

Integrity encompasses the following:

1. Ecosystem Health. This condition may apply to even non-pristine or somewhat degraded ecosystems, provided they function successfully as they presently are. Some examples might be (a) an organically cultivated farm, or a low input operation; but also (b) a lake which, having lost its larger species because of anthropogenic stress, now functions with a larger number of smaller, different species. Hence, ecosystems that are merely healthy, may encompass both desirable and undesirable possibilities, and may be more or less limited in the capacities they possess (or have become artificially or accidentally constrained by humans). It is for this reason that health alone is not sufficient.

2. Resistance to Disturbance. The ecosystem must retain the ability to deal with outside interference, and, if necessary, regenerate itself following upon it. This clause refers to the capacity to withstand stress. But non-anthropogenic stress is part of billions of years of systemic development. Anthropogenic stress, on the other hand, may be severely disruptive to the system in that it may contain realities that are radically new to the natural components of the system, or it may operate at intensities that are unprecedented in evolutionary history, hence the system may not have ways to correct or compensate for such stress.

3. Capacity for development. Integrity obtains, when at point C (see Fig. 1a, next page),[11] the system's optimum capacity for the greatest possible on-going developmental options within its time/location, remains undiminished. The greatest possible potentiality for options is also fostered by the greatest possible biodiversity (dependent on contextual natural constraints), as the latter is a necessary but not sufficient component of C. Biodiversity contributes to integrity in a least two ways: (a) through genetic potential, based on the size and diversity of population and their respective gene pools (hence supporting and enhancing both structure and function of these populations); (b) through biodiversity's dimensions as purveyor and locus of both relational information and communication, of which existing populations and ecosystems manifest and embody only a small proportion. We can only theorize about the immense capacities for diverse qualitative interactions among individuals and species, which are not presently existing or knowable.

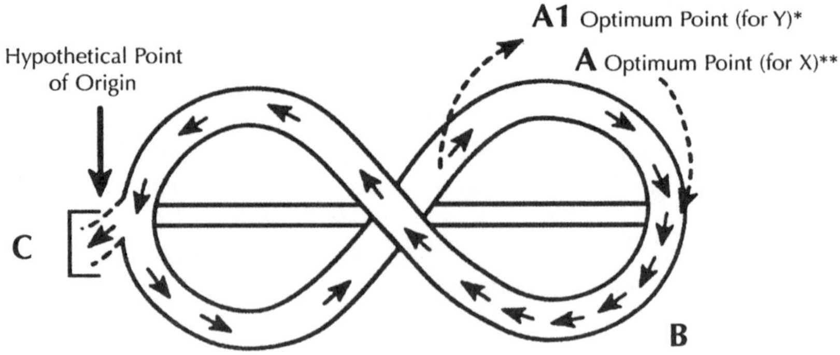

* e.g., deciduous forest ecosystem deliberately introduced
** e.g., naturally evolving Douglas fir ecosystem

FIGURE 1a (Hollings 1992). Arrows indicate directedness in the ongoing change and development of ecosystem.

4. Lack of Human Constraint. The system will possess integrity, if it retains the ability to continue its ongoing change and development, unconstrained by human interruptions past or present.

It is the latter two facets of integrity that differentiate it from the notion of ecosystem health. Although this paper is primarily concerned with "integrity" rather than "health," it is appropriate to show how it differs from integrity. What is the meaning of "health" in the context of an ecosystem? Is health to be classified simply as "the absence of disease"? Historically, once the "communicable diseases" were, to some extent, conquered in the biomedical model, noncommunicable diseases such as cancer, heart disease and other ailments of "middle and later life" became the main focus of study. Yet simple cause/effect relations are no longer sufficient to explain these diseases, nor to understand or even correct some of the complexities that lead to their occurrence. Breslow says, "the World Health Organization (WHO) has defined health as 'a state of complete physical, mental and social well-being and not merely the absence of disease or infirmity.'"[12]

Health has also been defined as the capacity to resist adverse environmental impacts at the present time, and as "the imputed capacity to perform tasks and roles adequately."[13] In this sense, "health" places "individuals on a continuum with physical, mental and social dimensions." This brief excursion into current "health paradigms" manifests a significant number of aspects of the concept which serve to separate it from that of "integrity," in the context of ecosystems.

The first difference is the most evident (keeping, however, in mind the limits of the single organism analogy): without one kidney, one may be healthy, albeit more prone to succumb to certain stresses and diseases; without one finger, one may be healthy, in the sense of being capable of resisting diseases as before the amputation. Yet in both cases, the organism's integrity has been diminished. This in turn diminishes either the capacity to withstand stress, or the potential for activities (e.g., one may probably not become a concert pianist with a missing finger). But there are many other differences as well, and perhaps a list of these differences would be useful to compare and contrast the two concepts:

(a) health in its more recent paradigms is understood as relative to population, social expectations and the like, rather than being defined substantially (as opposed to culturally); integrity, in contrast, is a real, actual condition (not a social construct);

(b) health is understood in relation to specific time-frames and locations; although the location of the ecosystem is determinant of its integrity (but not in relation to social standards);

(c) health is limited to actual manifestations in existing populations or species, instead of manifesting the less constrained, long-term intergenerational thrust of integrity;

(d) health strongly emphasizes functional over structural features;

(e) health is, at least partially, a social construct (both as phenomenon and as criterion);

(f) health, even "optimum health" may be understood with reference to "support/manipulation" (e.g., a "healthy" diabetic, permanently functioning on insulin).

Hence, the "ecosystem health" paradigm is a useful one, but it is applicable only to instrumental landscapes, defined within this work as possessing Ib. The fact that "ecosystem health" is compatible with (a) cultural relativity; (b) short-term, and a limited time-frame and scale; (c) present existing populations/species (vs. integrity's largely unpredictable genetic potential); (d) overemphasis on functional attributes; (e) social constructs; (f) support/manipulation—all separate it from integrity in its primary sense (as 1a), which is incompatible with all of the above. 1a is primary, with a meaning that goes beyond simple ecosystem health, which is 1b.[14]

Complex systems theory, in combination with its Aristotelian roots, helps to define and separate integrity (1a) from ecosystem health (1b). The latter, in essence, is compatible with external support and manipulation, as well as with the loss of some of its structural components, just as human health is. For example, without pushing too closely the analogy between an ecosystem and a human being, a person may live to a "healthy" old age, having lost a finger or a foot, or supported by insulin, as a diabetic. Healthy, manipulated ecosystems might be, for example, an ecosystem supporting organic agriculture, without toxic, intensive petrochemical inputs. But neither the best organically cultivated land nor respectfully treated areas devoted to sustainable forestry are "wild." For one reason (see "3. Capacity for ongoing development," above), ecosystem health (1b) is does not ensure the ongoing process of speciation: an ecosystem can be perfectly healthy, yet "the death of life" as evolutionary process will have occurred.

There are several areas in which Shepard's concerns and mine are coextensive, although, as we shall see, there are also some significant differences, particularly on the topic of "domestication" and agriculture. Still, the first point that must be clarified is that the principle of integrity (P1) is a principle designed to govern human interaction with the environment exclusively, not to prescribe for interhuman or social interaction as such.

Hence, I will not comment on any aspect of Shepard's platform that refers strictly to group interaction or social structures, although these, too, he argues, are drawn from our "wild genome." Ultimately, any aspect of technological modernity or post-modernity is relevant to ecology, as it influences our world view. Yet I will limit myself to those concerns which are more obviously ecological, rather than discuss the role of myths, or of education, or the question of "no history, progress or destiny," and the like.

The first point to note is that point "C," or the point of "optimum capacity" for an ecosystem (Def. no. 3 and Figure 1a), is an "optimum point" which is clearly limited by its time/location, as well as its structure and its imminent potentiality. It is not a point of "unlimited" choice and possibility. By acknowledging both structurally specific limitations, and the possibility of unfolding capacities that are not known at this time, the principle focuses on the primacy of the genetic make-up of both structure and function of the ecosystem, and recommends it as the goal and starting point of all ethical discourse. Hence appropriate "natural" ecolimits represent a clear and unequivocal starting point of moral considerations.

Another aspect of the definition of integrity that re-echoes the thrust of Shepard's argument, can be found in 3 (b), above. The role of the "wild" as a source of information and communication applies to all "individuals and species," including Homo sapiens. The "immense capacities for diverse qualitative interactions," however, are still to be understood as "interactions"and "capacities" within the normal potential of areas possessing Ia, or areas in the wild.

Hence the benchmark or standard function of the wild applies to all habitats, and all biota, including humans. Domestication as such is not the primary goal, but it is not always deemed to be evil, provided it is limited to "buffer" areas compatible with "core areas," or the wild. In "Post-Historic Primitivism," Shepard argues that when we don't attend to "human specific species-specific traits," we compound our problems, and court health disorders.[15] Shepard also advocates an absolutism parallel to that advocated by the PI: he shows that neither cultural relativism nor consumer preferences are sufficient to help us in our present predicament. Many of the "solutions" at which we arrive (in an attempt to resolve our health problems), such as a diet containing "organic vegetables and . . . alternatives to chemicalized nature," interaction with "small, face-to-face social groups whenever possible," the quest for "meditative stillness" in various forms, "regular exercise, especially jogging, . . . aerobics, [and] stretching" tend to "correlate with certain routines of life in cynegetic societies."[16]

Many of these aspects of our "wild" recovery are compatible with the principle of integrity, although it permits not only Ia or foundational, wild areas, but also healthy, "domesticated" buffers, or Ib. Finally, Shepard's position is an absolutist one. In a discussion of hunting, he says, "Neither the animal protectionists, the animal rights philosophers, nor the feminists hostile to vernacular gender, have ecosystems (including the wildness of humans) at heart."[17] I have argued elsewhere for a similar position, terming ecofeminism, for instance, an oxymoron. A combined platform including individual rights (feminism) and an ecocentric position, is self-contradictory, unless one takes the escape route of abandoning the quest for a moral position, for a purely situational, relativistic ethic.[18]

THE RECOVERY OF THE "WILD GENOME" AND THE QUEST FOR INTEGRITY: SOME DIFFERENCES IN A COMPATIBLE GOAL

What are the major differences between Shepard's position and the principle of integrity? The last quote in the previous section speaks of a problem related to hunting, and in Shepard's position on hunting and human diet, we find the major point of difference. The principle of integrity permits the adoption of what J. Baird Callicott terms the "three best ways" of eating:

> Purists like Leopold prefer, in his expression, to get their "meat from God," that is, to hunt and consume wildlife and to gather wild plant foods, and thus to live within the parameters of the aboriginal human ecological niche. Second best is eating from one's own orchard, garden, henhouse, pigpen, and barnyard. Third best is being or bartering organic foods from one's neighbors and friends.[19]

Reliance on chemical agriculture is not permitted, but domestication is (free-range and organic), of both plants and animals. My position is similar, although preference is given to forms of agriculture that are non-intensive, organic, respectful, and governed by Integrated Pest Management principles and to diets limited to fish and fowl, both free-range. This attitude is implicit in the acceptance of 1a but also of 1b, where health is compatible with supportive manipulation of the ecosystem.

In some sense Shepard is no doubt correct in saying that, at the deepest level, human "health" is to be sought primarily in our wild genetic makeup. Yet complex systems theory recognizes the ongoing developmental change of ecosystems (see Fig. 1a). Further, the second paragraph of the definition of integrity (see no. 3 (b) above) alludes to the fact that even the ongoing "figure eight" is insufficient in its one-dimensional representation to truly explain what happens as well as what may happen within an ecosystem's specific space and time. We refer to both interactive and developmental aspects which are neither present nor knowable at this time.

From that standpoint, I find it problematic to simply take a "wild essence" of humankind, without admitting the possibility of compatible changes in addition to or other than Shepard's own argument for hunting-gathering. Consider that the "hunt" as the sole or the main source of appropriate nourishment, is a practice that (a) no longer possesses its ceremonial and religious overtones, so that it is questionable whether it even retains its value in the present context, aside from diet questions; also (b) it is not available to the great majority of humans today, especially urban humans, but also human tribes in areas inhospitable to the small game that could conceivably nourish them (were they, say, to abandon herding, that is, survival based primarily on domesticated animals, as remains the case among some desert tribes in North and Central Africa). Further, (c) some "wild" landscapes have been affected by human abuse, for instance, through desertification, and it would be unrealistic to expect aboriginal lifestyles where "primitive land" (that is, unappropriated land in largely natural conditions) no longer exists.

Shepard might respond that (a) the ceremonial aspect of the hunt could be recovered, that (b) pastorality can be ecologically damaging to ecosystems, and that (c) he does not really advocate a "primitive" lifestyle.[20] Yet for now, I believe my criticism remains valid. For instance, I cannot see how camel milk or other herd-derived nourishment, could further harm the African deserts, where some of those practices occur; or, for that matter, what else, other than flown in foreign aid could possibly feed people in their native habitats, where "game" animals no longer exist. Yet for Shepard, "hunting-gathering" is more than a lifestyle, it is a "structure." As such, I cannot see any possibility of recovering it by either legislation or even consciousness raising (although some aspects, particularly as they relate to human health, can be regained through consciousness raising).

Aside from the myth and ceremonial motif, it is interesting to compare Shepard's position on human diet with that of Denis Burkitt, Ph.D. In a recent paper entitled "Are our commonest diseases preventable?", Burkitt argues that medicine is making "great progress," but that it is "headed in the wrong direction."[21] Rather than attempting to "mop up" the results of (a) an improper diet and (b) an inappropriate lifestyle, it should instead concentrate its efforts and research on discovering what constitutes an appropriate approach to (a) and (b). He also uses the exemplar of the "hunter/gatherer" cultures and compares the "modern western" diet (based on the changes occurring in the last 200-300 years, after the Industrial Revolution), with that of "Paleolithic and Third World humans," respectively. His analysis differs somewhat from that of Paul Shepard. (See Table 1).

TABLE 1 Estimated Energy Sources in Paleolithic, Third-World, and Modern Western Diets

	Paleolithic	**Third-World**	**Modern Western**
Starch	60%	60%	20%
Sugar	Minimal	Minimal	20%
Protein	20%	15%	15%
Fat	20%	15%	45%
	(Mostly saturated)	(Mostly saturated)	(Mostly Unsaturated)
Fiber	50-100 g/d	20-100 g/d	>20 g/d
Energy density	Dilute	Dilute	Concentrated

Burkitt argues convincingly that while our genes were capable of adapting to the major change in lifestyle with the "Agricultural Revolution, about 10,000 years ago," the modern Western diet instead is directly responsible for most "Western diseases" that medicine, as such, is impotent to correct (e.g., cancers). The similarity between "Paleolithic" and "Third World diets," is noteworthy, and clearly indicts our own rich consumerist lifestyle, and its failings. For instance, Burkitt cites Sir Richard Doll, in the *British Medical Journal*:

> Whether the object is to avoid cancers, coronary heart disease, hypertension, diabetes, diverticulitis, duodenal ulcer, or constipation, the type of diet that is least likely to cause disease is one that provides a high proportion of calories in whole grain cereals, vegetables and fruits; provides most of its animal protein in fish and poultry; limits the in-take of fats and, if oils are to be used gives preference to liquid vegetable oils; includes very few dairy products, eggs, and little refined sugar; and is sufficiently restricted in amount not to cause obesity.[22]

There is a marked difference between Burkitt's assessment of "improper diet," from a purely medical standpoint, although he accepts the desirable components of a hunter/gatherer lifestyle, and that described by Shepard. I believe it might be the former, rather than the latter, that is closer to what could and should be restored today.

Another practical problem is that there are some aspects of "culture" that even Shepard may not want to reject, such as universities to train and prepare persons such as himself, to research and teach, and doctors or other medical personnel to tend the many ills we and our children are presently prey to, at least before simpler modes of life can be restored—all require, minimally, the existence of areas where "culture" in the modern sense is permitted, provided it is not hazardous to core, that is, wild areas. If Shepard can teach well, and Dr. Burkitt can treat the sick, it seems counterproductive not to permit either of them to concentrate on their true vocation, leaving the "hunting" and the "gathering" to those whose particular talents don't match theirs. According to Shepard's argument, this would remain unfair to both scholars' true wild nature, but perhaps the communal aspects of their life would impose on them responsibilities for those who could not help themselves within their group (e.g., sick children).

Another difficulty is that the massive numbers of today's population, worldwide, require access to food sources that are reliably available and sufficient. Even the present national fragmentation is inimical to the fair distribution of food today, hence neither "hunting" nor "small groups" only would really solve the problem. Only a non-hazardous, sustainable global approach may work.

Hunting as such is not objectionable from the standpoint of integrity, but it creates other problems that must be faced. Most indigenous people have tended to overhunt before moving on, thus to effect ecosystems, habitats and populations of game animals. Their lifestyles are clearly not as destructive as Western technical assaults upon the wild, but their goals are not compatible with those of conservation biologists.[23] And if their goals and lifestyles are not compatible with conservation, then, in a world which is, on the whole, no longer biodiverse, yet where the "wild" is absolutely necessary for the reasons discussed, then the very lifestyle Shepard advocates would contrast with his own priorities, if practiced on a grand scale.

Finally, the viability of meat-eating today remains, for the most part, an open question, given the lack of religious and ceremonial aspects of the kill, and the lack of respectfully pursued and killed meat, in contrast with the availability of chemically grown and mass-produced meat. But the difference between Shepard and the vegetarians (and organic/free-range eaters) may be one of degree rather than of kind. Most vegetarians will accept that Inuit and aboriginal people may hunt and fish, but prefer for themselves organically grown food, some of them permitting some seafood and even fish. The latter is also permitted from the standpoint of integrity, as is free-range fowl. Aside from arguments taken from animal rights theorists, Shepard and ecosystem defenders ought to counsel the complete rejection of feedlot animals as food.

I agree that "we ourselves are genetically wild," but, as any wild creature, our habitat, to some extent, determines the form of our survival. It is right and appropriate to recognize the origin and the benchmark of what is appropriate for humankind. Yet it is equally necessary to recognize that an altered habitat coupled with our desire to retain some aspects of our modern culture, should lead to aware, but somehow modified practices. Our habitat, no longer pristine, will not permit a totally primitive lifestyle, nor will such a lifestyle support our immense numbers. "Limits" which are predicated upon non-existent landscapes and long-surpassed numbers of humans, would be as "bizarre" as the "social hallucination of no limits" Shepard decries.

Speaking of agriculture and the work of Wendell Berry, Shepard states that Berry's feelings about "fresh air and water, good soil and the sky, the rhythms of the earth and human sense in these things" were not "invented by farmers." Rather, "They are the heritage of the non-domesticated wild world. Much that is 'good' in his description does not derive from its husbandry, but from the residual 'wild nature.'"[24] It is not that "nature's order must be imposed from outside." Rather, nature's order must be protected from outside, especially today when, like a precious gemstone, it has become rare. Shepard says, speaking of feminists and animal rights defenders, that those whose main goal is the furthering of individual (or aggregate) rights, no matter how worthy, do not have "ecosystems at heart."

As for us, following the spirit rather than the letter of Paul Shepard's jolting and inspiring work, we might need to give primacy to wild ecosystems and landscapes, even at the cost of a less "wild" human existence. We may even need to prevent some tribal groups armed only with hunting knives (let alone those with snowmobiles and high-powered rifles) from hunting down and destroying the last of the "wild" beyond ourselves, even when this might be defensible (in an earlier world) from the perspective of restoring the "wild" within. That is the ultimate significance of the Principle of Integrity as well.

Notes

1. Paul Shepard, "Wilderness is Where My Genome Lives," 5th World Wilderness Congress (Tromsø, Norway: 1993), 3.
2. Laura Westra, *Environmental Proposal for Ethics: The Principle of Integrity* (Lanham, Md.: Rowman Littlefield, 1993), 3.
3. Shepard, "Wilderness," 2.
4. Ibid., 2.
5. Paul Shepard, "A Post-Historic Primitivism," in Max Oelschlaeger, ed., *The Wilderness Condition: Essays on Environment and Civilization* (San Francisco: Sierra Club Books, 1992), 48.
6. Ibid., 54.
7. Ibid., 55.
8. Shepard, "Wilderness," 6.
9. Ibid., 5.
10. Laura Westra, "Are Animal Ethicists Speciest?," Western APA Presentation (1993).
11. Figure 1a from C. S. Hollings, "Cross–Scale Morphology, Geometry, and Dynamics of Ecosystems," *Ecological Monographs* 62:447–502.
12. Lester Breslow, "Health Status Measurement in the Evaluation of Health Promotion," *Medical Care* 27:3 (1989), supplement S206.
13. Talcott Parsons, "Definitions of Health and Illness in the Light of American Values and Social Structure," in E. Gurly Jaco, ed., *Social Structure and Personality* (New York: Free Press of Glencoe, 1964).
14. See Westra, *Environmental Proposal*, for further discussion.
15. Paul Shepard, "Primitivism," 54–55.
16. Ibid., 55.
17. Ibid., 56.
18. See Westra, *Environmental Proposal.*
19. J. Baird Callicott, "Animal Liberation, A Triangular Affair?," in *In Defense of The Land Ethics* (Albany: SUNY Press, 1989), 36.
20. Paul Shepard, personal communication.
21. Denis Burkitt, M.D., "Are our commonest diseases preventable?," in *The Pharos of Alpha Omega Alpha* Winter, Vol. 54, No. 1(1991):19-21.
22. Burkitt, citing R. Doll, *British Medical Journal* 286 (1983):445-53).
23. Westra, *Environmental Proposal.*
24. Shepard, "Primitivism," 61.

Incorporating Bear[1]

David Clarke Burks

Thinking animals do not provide us with a literal model, but a poetic one.
—Paul Shepard, *Thinking Animals*

PLACEWAYS

ACQUIRING ECOLOGICAL WISDOM IS AN UNDERTAKING bound up with place. It grows with understanding that habitat in-forms. Knowing who we are is embedded in the soil of where we are. Ecology is household knowledge; knowing from where and from whom the honey comes.

Bear, especially for people of the northern hemisphere, has served over millennia as guide and talisman in the on-going search for Earth wisdom. Bear is connected, is integral. Bear knowledge can help us, as it did our ancestors, connect with the voice of Earth. Inhabiting place is knowing one's Bear-ings. Step in the tracks, not on the cracks. Follow all trails to the source.

Language, its origins and structures, is the story of a crossing into consciousness. Language is landscape-mapping in the mind through linkages of symbols connected at their roots. Following the pawprints of Bear, humankind throughout history has worked to map these connections. Using ritual, ceremony and science we forge an understanding of our place in communities and ecosystems. Totemic thought reaches across boundaries. As Paul Shepard writes in *The Tender Carnivore and The Sacred Game,* "Various plant and animal totems are not simply objects of reverence; they serve as codes for formulating and relating the conceptual systems of nature and culture.... In this way...animals and plants mediate between culture and nature in the mental life of man."[2]

From a study of some of the earliest Indo-European languages we learn that being in place is being one with the Earth. The Sanskrit word for Earth is *bhu*, a noun, which means the place of being. *Bhumi*, a cognate Sanskrit noun, signifies a particular location, ground, land or place. The verb stem, also bhu, means coming into being, or taking place. Being in place is coming into being. It is from this ancient Indo-European Sanskrit root that Anglo Saxons developed the word *beom* which signified a condition of beingness; I am. It is from the Anglo-Saxon that our English verb to be derives. Related, is the Teutonic stem word *ber,* which became the German *bar,* which is the origin of the word Bear in English. Being in place is bearing that place; putting on the robes of Bear.

Bear, Earth, being and place are interrelated in an ecology of consciousness which is seated in the web of Nature. Bear is a link to place, and place reveals being, the very ground of deep understanding. Standing under the constellations of the Great Bear, Ursa Major and Ursa Minor (the former known in Sanskrit as *rakh,* or "the bright") humankind pursues the polestar of understanding. The more we search the heavens and Earth the more we come to understand that all paths lead home.

Ursa Major and Ursa Minor are known in Navajo creation myth as Revolving Male and Revolving Female. We learn from Ray Williamson in *Living the Sky* that "the Navajo know it is time to plant when the constellation Revolving Male lies parallel to the horizon in early evening (late May or early June). In the spring, he is overhead; when he 'slants down to the east, it is winter.' The position of Revolving Male also indicates...when different animals are likely to mate, and when they will bear offspring."[3] Bear knowledge, like an aspen shoot, is connection to a vast network of roots and offshoots.

In *Thinking Animals,* Paul Shepard writes, "The only defense of non-human life not dependent on a technology that may change or an ethic that is largely idealistic is one close to human well-being, regardless of the changes in science and technology. It might be called minding animals."[4] The human mind needs Bear, and human intelligence is bound to the presence of animals. Shepard goes on, "They are the means by which cognition takes its first shape and they are the instruments for managing abstract ideas and qualities, therefore giving us consciousness."

Bear is a code image by which language retrieves ideas from memory at will. Bear is a means to self-identify, to become conscious of self by enabling us to objectify qualities and tracts. "By presenting us with related-otherness...they further...a refining and maturing knowledge of personal and human being." (*Thinking Animals*)

Bear inhabits the wild. It is the wild that homes original knowledge. Going home is going wild and getting acquainted with roots and berries. In the preservation of wildness we acknowledge the complex, generative forces that brought about our habitat. The work to be done is forest work, is mountain work, is coastal work, is the work of digging in. Working in place is performing original knowledge. As Gary Snyder writes, "Wildness is the state of complete awareness."

Chaos, collision, accident and design have all been at work since the beginning; are still at work, shaping our earth household. The vectors and tangents of creation, collapse, and rebirth have been traced out in intricate folds of the unconscious. The unconscious is collective, communal, ecological. We are woven inextricably into the fabric of place through ecological contents. Reinterpreting Jung's notion of the collective unconscious, Ted Roszak, in *The Voice of the Earth,* explains: the collective unconscious "might then be regarded as the repository of an evolutionary record that ties the psyche to the full sweep of cosmic history. Mind, far from being a belated and aberrant development in a universe of dead matter, connects with that universe as the latest emergent stage on its unfolding frontier."[5] The connection is integral. Bear is a link, sometimes referred to as an archetype, that associates the neolithic with modern consciousness. And the clasp, holding all the links together, is the ecology of place.

WORKING IN PLACE

Bear grandfather and Bear grandmother can teach us the old ways, which are new ways of understanding our place in this place we call home. Knowing where and how to gather huckleberries and boletus in the forests of Cascadia is one kind of work that connects. The fruiting bodies of place are a thousand local gods. Through them life becomes ripe. Through knowledge of them, through tasting them, we become placed.

Dis-covering the local gods is turning over new leaves; looking through surfaces, over and under rocks. Among the local gods in the Cascadian bio-region are: fungi, salmon, hazelnuts, marshes, camas roots, bearberries, gray squirrels, rivers, tree frogs, lichen, elk, cirrus clouds, soils, cattail and sedge, birds, rosehips, sunflowers, hemlocks, newts, damsel flies, smooth stones, waterfalls, human kindred, fir cones, meadows, sword ferns, lichen, voles, spores, seeds and nine hundred and sixty-nine times nine others. As Bear makes her rounds she is mind-full of all these beings. Following in her tracks we become Bear-full of all the constituents of place, and our native place in Earth's ever-unfolding geography.

Ancient people, indigenous people, have celebrated Bear through ritual and ceremony. Paul Shepard in his important work on bear, *Sacred Paw*, summarizes:

> Each one of these rites addresses a transformation, a process of change. In a world view in which man plays a cosmic participatory role, nothing is taken for granted. The rising of the sun, the turning of the seasons, the passages of a human life all require human action. But the basic process is always that of a wheel of existence whose movement is epitomized as a genetrix—the work of a metaphysical mother. A feminine principle of birth, growth, death, decay, and rebirth lies at the heart of the veneration of the bear, for the Bear is the supreme model—and therefore the guiding spirit—of the theme of renewal.[6]

Only four hundred generations ago we were hunters and gatherers. Less than 10,000 years is too short for evolution to uncoil from the double helix significant changes. As noted anthropologist Carleton Coon remarks, "We and our ancestors are the same people."[7] Knowing and praising the local gods is establishing connection with duration through place. Driving deeper, we are given to understand the very atoms of our bodies have been forged in stars, in distant nebulae. Could it be said that ontogeny recapitulates cosmology? Such a long round is the greater Gaian heritage.

THE WAYS OF PLACE

Along comes pigeon-toed Bear shambling through the underbrush of salal, making his way into a forest clearing. We are less than five hundred years removed from native Kalapooya hunters stalking, tracking every move and gesture of old grandfather-with-hair-on-his shoulders. Bear-teacher, bear-shaman in his travels reveals medicinal plants to cleanse the body, the location of springs to quench thirst, and what it is to walk slowly with great calm like a mountain. As Max Oelschlaeger reflects, "There is a pathless way for mountain walking."[8] And Gary Snyder advises, we must learn to walk "with the land and animals of Turtle Island in a sacred manner."[9]

A scent on a breeze yields a hundred whiffs of information. Ecological knowledge is understanding the unique character of all things and their native symbiosis. Understanding human connectedness with Nature is essential, as Ted Roszak notes, "to the health of both person and planet: the primitive, the organic, the feminine, the child-like, the wild." We must learn to follow our noses through the thickets. Paul Shepard reminds us in *Nature and Madness*, "There is a secret person undamaged in every individual...[and]...this means we have not lost and cannot lose, the genuine impulse. It awaits only...authentic expression."[10]

Learning to read the texture of soil, the stratification of rock, is learning the lengthy history of earth's becoming. Listening to stream music carries one through a round of the hydrologic cycle from the snow-covered peaks of the Three Sisters, down the chiseled valley of the McKenzie River, and on through flat coiling passages in the Willamette Valley until joining the Columbia River in its final run to the Pacific Ocean. Observing the character of place is the first step down the forest trail which leads to the waterfall. In the presence of the waterfall everything merges.

Language is a coding device for recall. What is recalled is attached to an image. Bear is metaphor, an analogue of relationships within ecosystems. Bear, in stories of indigenous people, is intuitive knowledge. As Shepard writes, "One may imagine speech in its primitive beginnings to have been entirely practical, a labeling of useful objects and acts. But it is just as likely to have been entirely composed of poetic terms with multiple meanings, what we now call metaphors, words connecting unlike things and events." (*Thinking Animals*)

Respecting the dignity of all things is recognizing intrinsic connection, a mutuality of interest among all things. An accord of dignity is an accord of peace which affirms the inherent worth of all parties. All things are not equal; rather all things are interdependent in community. Though unequal in weight, all things share inherent worth and deserve respect. Praising the local gods, the fruiting bodies of one's bio-region, affirms community in its broadest sense. Think locally, act locally, consume locally as Wendell Berry counsels. Living mycorrhizal is living in symbiotic relationship with roots of plants and animals and ancestors. This is the original meaning of ecological education.

As Bear walks beside a stream she listens to Salmon. Salmon is saying that water is dying. Dams and diversion channels have taken away water's ability to flow its native course. It is more than Salmon that is endangered, it is the very sculpting potential of natural processes. Dolores LaChapelle writes in *Earth Wisdom* that "The highest good is the optimally functioning ecosystem as a whole, not the immediate purposes of any one being in the ecosystem—man included."[11] She goes on to explain that living in balance with all beings is understanding the way patterns of energy move through place. Whatever we do uses energy from Nature; it is therefore incumbent upon us to return energy to Nature. To do this we must gather Earth wisdom by living in a manner that promotes diversity.

Embodying a "land ethic" is an acknowledgment that an action is right, in the words of Aldo Leopold, "when it tends to preserve the integrity, stability, and beauty of the biotic community. It is wrong when it tends otherwise."[12] Bear like man is both messenger and passenger. Bear is a litmus of the health of place. Displacing Bear is removing a way of understanding how we are doing, as well as what we are doing to Earth.

Bear can serve as tutor and teacher leading us over varied terrain to a more complete understanding of the principal rhythms in the annual round. Shepard concludes the chapter "That Ursine Quality" in *Sacred Paw,* with a story from which the following is an excerpt.

> So the Bear became the man's special guide in shaping all the separate acts of life into a harmonious whole, fostering in the man's mind the idea of coherence and sequence and even transcendence. Further, in its assured and solitary way the Bear seemed to know something that was not enacted, perhaps to have a secret. The man watched the Bear for many years and finally he understood: the bear, knowing what was to come from what was, was himself the eye of the future watching the present. His gift of revelation, unlike those skills which man could learn from the other animals once and for all, was continuous. His message was the health of the wilderness, which was always changing. The Bear was the voice of the earth itself. Then the man, remembering that the Bear was his mentor, realized that he himself had that voice, if only he could learn to sing as sweetly as the bear.

MIND-FULL

Bear-ing in mind is respecting the dignity of all things; the livingness of all things. Today our collective habitat is at risk because humankind's relationships with forests, streams, wild lands and wild lives are in distress. Being Bear-mind-full of place is leaving enough for succession. Changing the course of rivers, the biological make-up of forests, alters the ground rules. Only volcanoes can change the ground rules with impunity.

Knowing the right names, choosing the correct tools, overcoming ambivalence, these are some of the actions toward which Bear knowledge can lead us. Such knowledge will require transformations, changing shapes, changing our destructive ways to once again be in accord with the old ways. It is, as always, a matter of endings and beginnings. As noted in David Rockwell's *Giving Voice to Bear,* for the Tlingit people of the Pacific Northwest, *xuts tla,* or Bear Mother is the "Strong One" who nurtures and leads the way.[13] Listening to Bear is heeding the healing voice of the feminine.

Being Bear-ful in place is being aware of leaving enough for the next one who comes along. Clearcutting is destabilizing. Whenever diversity is removed and replaced with monoculture, whenever complexity is replaced by simple answers, the ability of habitat to restore itself is impaired. The more things become the same, the more vulnerable to manipulation they become. Diversity is the Earth's best defense against intellection. Straight-line intellection reduces things to commodities for the purpose of balance-sheet accounting. Once an ancient Douglas fir becomes a calculation of board feet it is reduced from guardian to victim, and so is the one who makes the calculation.

Knowing one's Bear-ings is adjusting one's behavior and demeanor to be alert to all the postures, odors, sounds and gestures in the surround. Assured-attentiveness and a keen awareness of intentions are beneficial sensors leading toward a deep awareness of place. The energy of place is inherent in existing footprints. Walking like Bear is following the path of native topography.

In his discussion of primal religions, Alan Drengson explains:

> The world is an ongoing, creative process to which we must pay close attention. We come to know it, not just by thinking about it, but by being immersed in it, by letting it speak to us, by being in it in a myriad of ways. So, be the tree, be the wolf, be the bear, be the flowers of the field and the mountains. They teach us to be free and whole and how to live with ecological wisdom.... Everything in nature is a teacher.[14]

As in forests so in streams, we arrive and depart with the ouzel, like John Muir becoming wet with the plasma of place. Deep communion with place occurs through immersion and emergence. I came back to myself as I was lost. Place-centering is healing the wounds of separation. Place-centering is taking strength from the wild. We become resource-full to the extent we become Bear-full. Bear resurrects life after a winter's hibernation. Spring bounds out of a winter's sleep reaffirming new life. Bear lives within us as we give voice to the earth. Incorporating Bear wisdom is pursuit of deep ecology. And once again Paul Shepard: "Like a ball at the top of a fountain, the human head pivots on its animal backbone, the mind a turning knot of thought and dream on the end of a liquid spear of living animals."

Notes

1. This is a revised and expanded version of an essay that originally appeared in *Wild Earth*, Vol.3, No.2., Summer, 1993.
2. Paul Shepard, *The Tender Carnivore and The Sacred Game* (New York: Charles Scribner's Sons, 1973).
3. Ray Williamson, *Living the Sky.*
4. Paul Shepard, *Thinking Animals: Animals and The Development of Human Intelligence* (New York: Viking, 1978).
5. Theodore Roszak. *The Voice of the Earth* (New York: Simon & Schuster, 1992).
6. Paul Shepard and Barry Sanders, *The Sacred Paw* (New York: Viking, 1985).
7. Carleton S. Coon, *The Hunting Peoples* (Boston: Little Brown and Co., 1971).
8. Max Oelschlaeger, "Mountains that Walk," *Wild Earth*, Vol. 2 No. 3, Fall, 1992.
9. Gary Snyder, *The Old Ways* (San Francisco: City Lights Books, 1977).
10. Paul Shepard, *Nature and Madness* (San Francisco: Sierra Club Books, 1982).
11. Dolores LaChapelle, *Earth Wisdom* (Silverton, CO: Finn Hill Arts, 1978).
12. Aldo Leopold, *A Sand County Almanac: With Essays on Conservation from Round River* (New York: Random House, 1970).
13. David Rockwell, *Giving Voice to Bear* (Niwot, CO: Roberts Rinehart, 1991).
14. Alan R. Drengson, "Paganism, Nature and Deep Ecology," *The Trumpeter*, Vol. 5, No. 1, Winter, 1988.

Returning Home: A Journey With Paul Shepard

Kraig Klungness

MID-EVENING, JUNE, 1991. Sitting, reading, occasionally glancing to my right, through the north window, beyond the white birch and spruce, to Lake Superior. It is very calm, mirror-like, inviting. There is still time, with the days so long, for a quiet paddle of the kayak along the red sandstone shore (Northwoods Slickrock). It is a beautiful shore, heavily wooded above the sandstone cliffs from which small streams cascade directly into the Lake. Exposed roots of birch and mountain ash grip the cliff's edge in anticipation of a fateful plunge into autumn's rough seas. For many miles, this shore is wild.

I concentrate on finishing a passage before departing with the kayak. A soft thump sounds from the steps to my door, just outside the south window. Probably a red squirrel, I think, without looking up. Then another thump and the window light dims. My flesh tingles mildly as I slowly turn my head for a look. Standing on hind legs, three hundred pounds I would guess, is a magnificent black bear, shiny fur, paws on the window, nose almost touching the glass.

Slowly I rise from the chair, straighten up and turn toward the window. The bear remains, our heads at the same height. Silently sliding my stocking feet in the bear's direction, I too lean into the glass until my nose almost touches, gazing straight into its small, curious, brown eyes. Ten seconds, fifteen seconds perhaps. Its head sways, nose sampling air. Then gracefully, quietly, it puts all four paws to earth and ambles into the forest.

~ ~ ~

Utah State University, 1977. The madness never stops, I think. Can it stop? Can humans, can I, ever know wholeness? Ever again live with, rather than against, nature and wildness?

My desire to defend the wild goes back to growing up at the edge of town, exploring the neighboring forest; and boyhood hunts with patient grandfathers. Now in college, I hunger for the connection between the natural world and a healthy psyche. I agree with Gregory Bateson's speculation that industrial humans are living primarily within a laboratory of their own creation, and, as he asserts, only in a laboratory context will a wild animal go insane in response to random rewards. Wildness embraces the chance gifts of a congenial natural world, the human engineered laboratory does not. If you drive the natural world insane, says Bateson, "its insanity is incorporated in the larger system of your thought and experience."[1] We are driving the natural world, and therefore ourselves, insane.

How to heal? I want to know. A valued teacher and friend casually mentions Paul Shepard's *The Tender Carnivore and the Sacred Game* while attending to my speculations. I buy it, read it, and begin to share an incisive but gentle scholarly journey into our past, into our psyche, and out to biotic community. I then read *Man in the Landscape*, and, in following years, *Nature and Madness* and *The Sacred Paw*. In Shepard's work I find an amazingly deft synthesis of biology, anthropology, art history, and psychology that provides a holistic evolutionary context for the defense of nature and wildness, and the healing of psyche.

> It is convenient to blame the destruction of nature on impersonal, inexorable, collective forces incidental to progress, technology, or civilization....The only certainty we have is that it is our monster; it is us.[2]

As biocentric wilderness activists, we fight to save habitat for nonhuman life, measuring our success in habitat preserved and species saved from the tidal wave of anthropogenic extinction. Our efforts for Other are inseparable from our own inner being. With each extirpation, with each wild acre lost, we are personally diminished, injured, damaged, and the possibilities for healing human dis-ease, less.

The sadness of this loss is oceanic. Our organic embeddedness is real. In 1991, when I looked into that bear's eyes, I felt a shared fate and a shared meaning. To look into bear's eyes was to look into my own. To be whole, wilderness activism needs a vision of human existence that reflects this connection, honoring free nature.

My original sense of biocentric connectivity was intuitive. Much of the clarification and broader understanding came from Paul Shepard's work, including explication of my underlying intuition as genetically encoded. From him I know that the nature hating will not stop until our own psychogenesis is whole. And I know that biophilia is instinctive, not a fantasy, which provides a place for hope.

To discuss what Shepard's work means to me as a wilderness activist, I need to share the personal journey his writings have provided, and their bearing on recovering wilderness as home.

BIOPHILIA

> ...well-being does not demand the sacrifice of the beauty of the world, which is the esthetic experience of its physiology.[3]

Wild nature shaped us, honing our body, our nervous system, and psyche in beautiful coevolution with all of life. I love nature's complexity, its texture, moistness, smells, sounds, silence, dark recesses, bright exposures, the seen and unseen, its mysterious presence. From Shepard comes the beautiful descriptions of why my being responds as it does, of human biophilia, the biology of our eyes, the phylogeny of our psyches.

> Human culture and life are inseparably locked into organic interrelations, element cycles, and energy transfers that bind many species of plants and animals to place and time and each other. This brotherhood has two dimensions; ecological, the dynamic web of the living; and evolutionary, the converging lines of antecedence. There is no freedom to be found in its denial[4]
>
> The human—the vertebrate—eye originated in the sea. Its basic structure is the same for all vertebrates, the most primitive of which are aquatic. . . . Compared to eyes independently invented in the sea, such as those of the squids, the similarities are astonishing.... With some slight additions and modifications,...the vertebrates carried this eye onto the land.... Our thoughts are teased by the consequences of a rebuilt sea eye on the delicate ego of a grounded primate, or, more exactly, the role of a salvaged reptilian-amphibian-ichthyoform eye in the making of man.[5]
>
> Nature is real and love of nature is part of its reality.[6]
>
> So we are taken on an evolutionary journey, almost animistic.... The living cell has its foundations deep in the physical structure of the world, where boundaries between life and death are indefinite.[7]
>
> [E]mbracing human biology while preserving our animality...What they [humans] must learn is to be an [wild] animal. If they learn otherwise it may kill them and life on the planet.... By acknowledging what is shared—the only secure grounds on which identity can grow—we would find ourselves.[8]
>
> [W]ith little patience for human chauvinism...man is not the end all of all evolutionary change.... It is both erroneous and wrong philosophically, I think, to assume that all creatures want to be men or that human consciousness is the cosmic pass toward and through which all life must make its way.[9]

Even in his first book—*Man in the Landscape* (1967)—he was a biocentrist and ahead of his time.

The importance of taxonomic experience and therefore biotic complexity to our psychogenesis makes clear the continuity of our intelligence with that immanent in nature. Nature's mosaic is kaleidoscopic, as is the primal mind of hunter-gatherers. Their culture was referential to Other rather than self, plant and animal totemic, and diffuse with the diverse patterns of natural process. Self-understanding and personal and social integrity were the allowances of ecological integrity. Sensibilities essential to the hunt, and interaction with death, engendered ecological compassion.

The prevalent historical attitude that culture is unique to humans, exempting us from animality, is an anthropocentric fantasy. Culture, acquired through capacity for language (sign or symbol), is as embedded in biology as instinct and, as Shepard points out, occurs in many animal species.

This has recently received much attention in the popular media, but the scientific findings are not new. They have been repressed or ignored by a society fearful of what an etiquette that equally embraces nonhuman species will do to its biocidal economic agenda.

In his argument for the commonality of culture across species in *Man in the Landscape*, Shepard discusses linguistic studies finding homology between the language of bees, dolphins, and humans. He expresses his perspective on human creativity within this biological context, never letting us forget our phylogeny:

> ...but our experience has not been only that of humans. We and the painter chimpanzees of England share a delight in abstract color schemes, curves and vertical and horizontal lines, a mutual sentiment for the forest —a heritage from mutual ancestors. Man does not have a priority on creativity, which is a pervasive activity of life, and perhaps of the universe.[10]

As culture is a biological commitment, so is low population density. Trophically omnivorous, but socially a carnivore, with humans, as with all large carnivores, low birth rates and low population densities are the preferred survival strategy, our natural state. The human herd, dependent on domestication, is a bizarre aberration from our animal being, destructive of our own and Other's ecology.

> No one thinks of a herd of bears, a drove of tigers, or a flock of eagles.[11]

The prudent predator is not an exterminator.[12] To love nature, we must limit our reproduction.

In biophilia we find our indigeneity, our place in nature, psychogenetically, ecologically, temporally. But more important, we feel the intrinsic rights of natural diversity, of all species and habitats, to exist. That the value of biotic complexity is intrinsic, and does not need to be defined through me or through us seems encoded in my cells, all cells, and communicated from them.

ALIENATION

> Genocide began on earth with the extermination of hunter-gatherers and extended as biocide to the selective destruction of wild plants and animals on the new basis of politically structured values, a psychological

extension of the tribal wars of civilization.[13]

Our ecopathology began with the domestication of plants and animals, the first gestures at managing nature through reduction, giving rise to agriculture and laying the foundation for Cartesian thought and the machine age. This was not something most humans lusted for. Its genesis was slow and complex and the reasons are still debated, but it is known that then and now primal peoples around the Earth resisted and are resisting its encroachment, albeit unsuccessfully. "God damn a potato" quotes Shepard of Chief Washakie of the Shoshones in "A Post-Historic Primitivism."[14] "The invention of drudgery and catastrophe" is a description he gives of agriculture in *The Tender Carnivore and the Sacred Game*.[15]

Once in motion, the momentum and inertia of this ecopathologic transformer of landscapes and mindscapes into infantile "bovine sanctuaries," has been unstoppable, smashing its way through ecosystems and minds amidst a torrent of ideological rationale worshiped in flaccid, domestic docility.

After agriculture began, there was "lust for more children" to be used as farm implements.[16] Obsession with production paralleled obsession with reproduction, reducing women to "the job of emulating barnyard animals, then of becoming baby machines."[17] Land became an instrument of production, a possession, the world was presumed to be owned by men, wild became the enemy of the tame, the one God our father, we his sheep, and wilderness was no longer home. With the amassing of food, children, territory, and animals, toil was sacralized, leisure trivialized to idleness, and dreamtime was driven to unconsciousness. Theocratic state power grew along with trophic anxiety. Shepard emphasizes that it was never a matter of whether the crops would fail, but a matter of when.

The integrity of diversity eroded in Other and in us as psychogenesis was mutilated. Monoculture expanded across Earth with the march of the self-referential and self-maximizing entities glorified in the ideology of history. And history is what we must leave behind to turn away from the failure of agriculture. It is a linear temporal process hostile to habitat that insists we cannot recapture our deep past.

Even adept critics of industriarchy are unable to venture beyond romanticizing the family farm as a return to nature and spiritual wholeness. Shepard skillfully debunks the arguments of this genre's most skillful advocate, Wendell Berry, in "A Post Historic-Primitivism":

> We may ask whether there are not hidden imperatives in the books of Wendell Berry obscured by the portrayal of the moral quality, stewardship syndrome, and natural satisfactions of farm life. He seems to make the garden and barnyard equivalent to morality and esthetics.... [T]here are reasons to wonder whether the metaphors that mirror agri-culture are not infantile. (For hunter-gatherers the living metaphor is other species, for farmers it is mother, for pastoralists the father, for urban peoples it has become the machine.)
>
> ...Berry writes with great feeling about fresh air and water, good soil, the

> sky, the rhythms of the earth, and human sense in these things. But those were not invented by farmers. They are the heritage of the non-domesticated world. Much that is "good" in his descriptions does not derive from its husbandry but from the residual "wild" nature. He accepts Biblical admonishments about being God's steward, responsible for the care of the earth...another way of saying that the world biomes need to be ruled, that nature's order must be imposed from outside.[18]

THE RECOVERY OF WILDERNESS

> Science now hangs a radio transmitter on the bear's neck to trace his movements, but we may ask who is really lost?[19]

Lake Superior was still serene as I slid the kayak off the wet, slick sandstone, into the water, imagining where in the landscape the bear of that evening might be. I shuddered to think what this world would be without bear. Throughout the Earth, human agency was, with some exceptions, depriving bears of habitat and health, causing their decline.

What would this world be if not graced by the more-than-human? A hundred species a day leave this Earth, their voices not heard, their inherent right to exist overruled. Without them, where are we? What are we?

Those who romanticize the agricultural life fail to realize that it, not primitivism, is the dissenting tradition from human ecology. With it, we went from wilderness as home, wildness as sacred, to seeing wilderness as disorder, chaos, evil, and raw material for progress. We literally left home, abstracting ourselves into a virtual reality of domestic "genetic goofies," stunted psyches, bodily displeasure, extirpation of wild, and resourcism. Individual identity suffered accordingly.

> For such an individual the nonhuman elements of the self must be regarded with suspicion. One's bowels, like snakes or grasshoppers, must be regulated. One must take charge of the inner and outer world, either as a tyrant and spoiler or as God's steward.[20]

But we have in our species, says Shepard, the "call of the Wild," as do all wild species. We need a home that calls forth this wildness; a spacious existence where identity is inseparable from place; a gentle and graceful ecological life. Put the boundaries around civilization, contain it. At this place in time, wilderness recovery is a basic need.

Shepard's criticism of the scenic wilderness esthetic does not leave wilderness activists without a fundamental concept. It leads us to the idea of wilderness as vast, unmanaged habitat, existing for its own sake.

Until the more recent biocentric defense of wild lands, wilderness had been subverted into a fashionable sentiment amidst industrial excess—a scene of human design extending from a Cartesian sense of nature being external and separate.

> If, basically, the landscape meant the wonder and mystery of the natural world, those feelings of an intuitive sense of shared existence and a common ground of being, it was used to subvert and capture that impulse by transforming it into scenery. Thus it served corporate greed and the industrialization of the world.[21]

Conceding the token preservation of grand scenery, "wise use" continued across most of this continent, putting fragments of nature on reservations, exploiting the rest. It was our manifest economic destiny. The natural world was nothing but food for a march toward some distant, undefined utopia where one never arrives. The national parks we value so much were established from this ethos.

> . . . . as the multitude was vomited into the edge of the primeval wilderness by democratic travel a few salvaged perceptual trophies of great value, which would lead later to the extraordinary founding of the national parks.[22]

Shepard gives life to the deep intuition that Earth itself must remain wild in reciprocal relationship to our own, and Other's, wildness. And wildness does not flourish when confined, in the landscape, or in our own bodies and minds. Our relationship with wilderness, to be animate, needs to be less abstract and more engaging of participation in natural process. It cannot be in conflict with wildness. Wilderness cannot be a management zone; an extension of resourcism, which Shepard has defined as "the most insidious form of nature hating because it poses as a virtue, as prudent, foreseeing, and unselfish."[23]

As an extension of the need to love and be loved by nature, wilderness becomes a concept of comprehensive healing, arising from the ecology of ideas as an antibody to our 12,000 year-old ecopathology. And while it can be argued that the idea of wilderness arose from the ideology of exploitation, all healing agents arise from the stimulus of disease.

To be healing, wilderness must be vast, have connectivity, its value intrinsic. The vastness naturally extends from two essential principles: that the wilderness exists as an evolutionary context for free natural process and species indigeneity; and, that it allows the complexity of natural experience essential to an ecological psychogenesis—a psychogenetic maturity that leaves nature-hating behind.

In the mosaic of big wilderness there is a biological richness and abundance that once again gives meaning to leisure and puts wildness back into cosmology, myth, and psyche. The self-patterning of free natural process can metaphorically repattern our own minds away from management, toward an ethic of letting things be; of witnessing, going with, but not controlling, nature; of being more than self-referential. That wilderness be vast is essential to its rightful place in our consciousness, to an ecological coupling with our genome, and to the needs of natural diversity.

CONCLUSION

To primal peoples of the northern world, the body of bear, slain in a holy hunt, was felt and seen, as stated in *The Sacred Paw*, "as a message of good ecology and good relations between them [the people] and the sacred powers that govern the world."[24]

As an activist, I need context for the possibility of not being lost. Through Paul Shepard's work I have gained this, and I thank him, and I thank bear.

The deep past is here,
accessible.
Some find it.
Wildness begins to aikido the
machines.

Time spirals as history dissolves.

Wilderness returns as home.

Notes

1. Gregory Bateson, *Steps to an Ecology of Mind* (New York: Ballantine Books, Inc., 1972), 484.
2. Paul Shepard, *Man in the Landscape*, 2d ed. (College Station: Texas A & M University Press, 1991), 214-215.
3. Shepard, *Landscape*, 274.
4. Shepard, "Introduction," Landscape, xxxvi.
5. Shepard, *Landscape*, 3-4.
6. Shepard, "Introduction," *Landscape*, xxxviii.
7. Paul Shepard, *The Tender Carnivore and the Sacred Game* (New York: Charles Scribner's Sons), 25.
8. Shepard, *Tender Carnivore*, 83-84.
9. Ibid., 102-103.
10. Shepard, *Landscape*, 23.
11. Shepard, *Tender Carnivore*, 97.
12. Ibid., 127.
13. Ibid. 35.
14. Paul Shepard, "A Post-Historic Primitivism," in *The Wilderness Condition: Essays on Environment and Civilization*, ed. Max Oelschlaeger (San Francisco: Sierra Club Books, 1992), 60.
15. Shepard, *Tender Carnivore*, 16.
16. Ibid., 244.
17. Ibid., 244.
18. Shepard, "Primitivism," 60-61.
19. Paul Shepard, "Introduction," *The Sacred Paw* (New York: Viking Penguin Inc., 1985), xiii.
20. Paul Shepard, *Nature and Madness* (San Francisco: Sierra Club Books, 1982), 105.
21. Shepard, "Introduction," *Landscape*, xxv.
22. Shepard, *Landscape*, 151-152.
23. Ibid., 236.
24. Shepard, "Introduction," *Sacred Paw*, xviii.

The Migration of Eels and Other Observations About Renewal

Michael Martin McCarthy

For a time I lived in Australia at a property named "Watersong." The entire northern and eastern boundary of this forty-five acre parcel was defined by a mile's length of Jackson's Creek, that tumbled over a series of waterfalls before reaching the sea, fifty miles away.

Jackson's Creek is in a grassland—a dissected plain. It is a river in a canyon, with the dominating aspect of the landscape created from lava flows thousands of years ago. This surface of igneous rock formed box canyon conditions and step-stone bars across the river with continuous waterfalls and small pools. The largest waterfall on the property fell nine feet into a pool two hundred feet long. This pool, below the house, was surrounded by high and sheer canyon walls. The canyon land river corridor, with intermittent waterfalls, was as diverse with animals and plants as the plains above were redundantly simple.

There is a sequence along the river. It begins in the Australian spring, in August, when the Ravens (*Corvus coronoides*) build their nests. A few weeks later, the yellow blossoms of the Blackwood (*Acacia melanoxylon*) trees create a scene up and down the riverway that lets you identify every single one within your view. By November, Black Wattles (*Acacia mearnsii*) are blooming pale gold; and in December, the Wooly Tea Trees (*Leptospermum lanigerum*), with their grey green leaves and white crab-apple-like flowers, fill the edges of the river channel wherever the current is not too fast. As summer extends itself across the grassland the phenology of the river landscape contracts, or so I thought.

It was mid-January of 1987, while making my way along the river, when I first saw the eels (*Anguilla australis*) ascending a waterfall. I remember the captivating attention as I saw groups of twenty, then realized that there were hundreds and then, as I looked up the river, saw that it was moving with thousands of small eels swimming upstream, against the current. Uncountable squadrons of tiny eels, three to five inches long and lucent olive brown were wiggling up the face of the waterfall in front of me wherever it was moist. I stood, enchanted.

The eels were congregated in the pool below the waterfall and with a slithering, climbing motion moved up every available inch of rock surface on either side of the rushing water. Many slipped to the water below from the force of the fast current but the furious migration kept coming. As I moved upstream past the largest waterfall, I could see that the eels had surmounted each of the seven falls on the property with intense purpose.

These eels are primarily nocturnal in fresh water and what I saw that afternoon may have been far less than the migration that occurred in the evening. By the next afternoon the numbers were reduced to one tenth of the first day and on the next, they had disappeared. Adult eels spawn in the Coral Sea off Queensland, Australia about 2,000 miles from the waterfall where I stood. The eggs are carried southward by the Australian current and after metamorphosing into unpigmented glass eels, they enter estuaries and wait until they migrate to the upstream reaches of coastal rivers.

Of course there is also a downstream migration of the adult eels (they may take ten to twenty years to mature and reach a length of five feet) but the young eels are of more interest because of their incredible numbers and because young eels are delicious, especially when smoked. Torch lights reflecting off the waterfall walls would have made catching eels easy for the presettlement aboriginal tribes. Up on the higher plains, at seasonal camp sites, the fires for smoking would have lit the entire river corridor.

According to Australian archeologist Josephine Flood, there is considerable evidence of specialized eel traps and other structures used to harvest eels on a seasonal basis.[1] She described the diaries of G. A. Robinson who in the early 1840s, did sketches of eel pots, eel weirs and sticks upon which captured ells were threaded. These archeological records confirm a semi-sedentary way of life, that began thousands of years ago. I often sat on a bluff overlooking and across the river from my property and there, overgrown with shrubs, was a semi-circular, three foot high stone wall, that I imagined was a remainder of one of the seasonal eel camps.

While concentrations of young eels could be harvested from the waterfall rock faces in Jackson Creek during early summer, there would be other animals and plants to collect when that migration was over. The Bogong (*Agrotis infusa*) moth migrates in the millions to the high altitudes of the Snowy Mountains. Josephine Flood points out that their main value to the aborigines was the ease in harvesting, because hundreds of moths lie side by side on each rock face. Scraping a stick along the rock makes the moths fall into a waiting container, after which they could be cooked or smoked. This was considered a rich and excellent food source.

Neither eel gathering or moth collecting provide much evidence for the archeological record. It is even more difficult to determine if the migrations enabled large meetings of different tribes or if small groups kept to certain territories. But it seems plausible that these hunting-gathering people had developed a complex and rich culture around certain biological events that occurred with regularity, year after year. It was part of their renewal.

I had not seen the eel migrations when Paul Shepard and I sat at the base of that waterfall a few years earlier. We had talked about renewal and human ecology—a subject we often discussed, and Paul brought new insights because that was the year *The Sacred Paw* was published. I remember Paul saying: "From the beginning, our ancient myths, have been about renewal— whether it was reincarnation, food replenishing, initiation rites or as a model for immortality. The stories are among the earliest records of human intelligence—the sagas of the annual cycles of the seasons and the concept of time."

As we sat talking about natural cycles, the Brown trout (*Salmo trutta*) in the river were making surface indentions as they fed on an emerging hatch of mayflies. The trout (now a self-sustaining population) had come from Europe by way of Tasmania, whose Premier, Sir James Willson Agnew, had arranged for the over-sea transplanting in the mid 1800s.[2]

It seemed so like all my discussions with Paul—intense and intellectually exciting—sitting on the river bank, watching an introduced European trout feed on mayflies indigenous to Australia, both following natural cycles and patterns that were thousands of years old. And during all this time, we continued to talk about cave bears killed by Neanderthal men, dated by carbon-14 testing at nearly 50,000 years, and their myths about renewal.

The sense of renewal is a theme in all cultures. When the geese return in spring, when the salmon migration moves upstream, when caribou move across autumn's tundra, and when eels migrate, these are symbols for our own renewal. It tells us that once again, for each there is a season and that everything is reborn. Places that have events that are seasonal become special places. We want to become caught up in that phenomena, in that sense of wonder and sense of place. The sense of renewal which comes from understanding nature's cycles is an essential part of understanding our own human ecology.

This need is rooted in the hunting-gathering traditions that we evolved with for so many years. The hunter must understand and must develop passionate attention to natural cycles. The process places him back at the same place for the next harvesting of the seasonal bounty. The eel I harvest this year returns to me next year. The eel becomes part of the hunter's world, not isolated from it.

In his study of Eskimo cultures in the Arctic, Barry Lopez commented about our civilized way of making everything an object: "A fundamental difference between our culture and Eskimo culture, which can be felt even today in certain situations, is that we have irrevocably separated ourselves from the world that animals occupy. We have turned all animals and elements of this natural world into objects."[3]

A few years ago I was in Jackson Hole, Wyoming. It was about half past seven, a pleasant autumn evening highlighted by a nearly full moon. As I walked among other window shoppers, tourists and residents, I heard the sound of a flock of Canada Geese (*Bronta canadensis*). Looking up, I could see the "V" pattern over the mountains, crossing the moon's face. Flock after flock passed over, the kind of mass migration that signals an approaching cold front. This spectacle of migration went on for some time and, while I was standing in the middle of the walkway gazing upward, a young man accidentally bumped into me. He was wearing a "Save the Whales" shirt. Looking around me, I saw that no one was listening to the geese, no pair of eyes looked to the evening sky. Yet, those people visiting and living in Jackson Hole probably represented a more environmentally concerned group of people than exists in most communities. As the geese disappeared, another person passed with a shirt that pleaded to "Save the Rain Forests." I want the rain forests and the whales saved too, but it is useless as long as they are only objects. Soon some other objects and other forms of entertainment will catch our fickle attention. We need a much deeper relationship with the natural world and its rhythms—something that approaches the sort of hunting/gathering economy we evolved from, and we need it for our own good, to renew us. Others have written eloquently about this, for example: Roderick Haig-Brown, describing his needs in the modern world:

> I want the seasons to have full meaning for me; I want to know storm and fine weather and to have to be out in both; I want a river within sight and sound of me, the sea and hills within reach.... I want to continue as a man not too remote from the sensations and thoughts of my forebears.[4]

Laurens Van der Post, discussing bushmen of Africa:

> The essence of his being, I believe, was his sense of belonging: belonging to nature, the universe, life and his own humanity. He had committed himself utterly to nature as a fish to the sea. This feeling of belonging set him apart from us on the far side of the deepest divide in the human spirit.[5]

Paul Shepard, describing a different history about the barbarians from the north:

> ...[T]he plundering Norsemen, Goths, Vandals and pillaging Vikings—are but distortions of traditions that earlier had a different tenor, often expressed as a subtle and beautiful collaboration between creative men and the natural world around them.... The older pagan thought was a culture in many ways as spiritual, elegant and intelligent as that shaped by men anywhere. At the very least it cherished the earth and all its life. At its best it may have been a true source of civilization's most valued ideas.[6]

I always thought of Paul as being on the frontier; that his work was scouting the boundaries of human ecology. He offered a glimpse towards a different, more related world, beyond simple appreciation. Although I never had a class with Paul, his friendship and his writings provided the insight to pursue the newly merged fields of human ecology and environmental planning.

Years ago, Paul asked a simple but difficult question in a class I was teaching. The class was about environmental planning and development. Paul asked: "How much should human ecology or an evolutionary perspective inform and affect the concerns of architects, planners, builders or developers?" This is a very good question and hits at the heart of what is for the most part, an abdication by those professions.

My education is as a planner and designer. I have worked in corporate and academic settings. My experiences are broad, but my hopes for these professions that design, construct, and develop our environment are modest. In its aggregate, this industry represents over nine percent of the gross national product of the world's leading nations. This is the industry that immeasurably and permanently changes and modifies the environments in which we live: our neighborhoods, our towns and our wilderness. But with few exceptions, Paul Shepard's question remains unanswered. Human ecology does not inform the practice of these professions. A concept like human renewal or designing for human well-being is foreign. Few wonder about the migration of eels or any other natural cycles.

The United States National Environmental Policy Act (NEPA) has had far reaching effects on the development and construction industry, although often not as much as its proponents might have wished. According to Valerie Fogelman, the late Senator Henry Jackson was author of another part of NEPA that never got out of committee. If enacted, it would have made the planning, design and building industries much more responsive to human ecology.[7] Jackson had wanted to establish everyone's inalienable right to a healthful environment. The critical word is healthful—an optimum human environment and sustaining natural world. A similar, new initiative is needed today. Could there be any more important amendment to our Bill of Rights?

Being part of and understanding natural cycles is necessary to any optimum, healthful, human environment. There is a surge of renewal released from this deeper interaction. We are renewed whenever we reaffirm our connection to this cyclical world. These seasonal phenomena occur everywhere. I am not arguing about the importance of some spectacle of wildlife that only happens in wilderness areas. The eel migration I saw occurred within thirty-five miles of a metropolitan area in Australia with a population of three million people. Today, every autumn and spring I watch for the thousands of geese that fill the fields within an hour's drive from one of Texas' most populated metroplexes. These natural cycles can be found, when we look. But we are so out of touch from non-use with the rhythms of nature that our senses have left us. We are the poorer for that and there is considerable evidence that our own health is impaired and that ultimately our immune system is affected. Our lives and their quality could be enriched and renewed. The knowing of place and season and being in an interlocked web with these phenomena— appreciating their permanence over all time and their impermanence until the next cycle, are wonderful, uplifting thoughts.

Human well being and environmental development are interwoven. Our lives are journeys enriched by our connection to the renewing rhythms and cycles of the natural world. Over the years I have prospered by understanding natural cycles, by becoming interlaced with these everlasting yet ephemeral phenomena and thereby feel a sense of renewal. I have become healthier from watching the migrations of eels and the flights of Bogong moths.

Notes

1. Josephine Flood, *Archaeology of the Dreamtime* (Sidney, Australia: Collins, 1983).
2. Lynne Agnew, of Wavery Croft Station outside Oatlands, Tasmania; personal conversation about family history.
3. Barry H. Lopez, *Arctic Dreams* (New York: Charles Scribners' Sons, 1986).
4. Roderick L. Haig-Brown, *Measure of the Year* (New York: Lyons & Burford, 1990).
5. Laurens Van der Post, and Jane Taylor, *Testament to the Bushmen* (New York: Viking, 1984).
6. Paul Shepard, *Man in the Landscape* (College Station: Texas A&M University Press, 1991).
7. Valerie M. Fogleman, "Towards Stronger National Policy on Environment" *FORUM: For Applied Research and Public Policy* 8, 2 (1993).

The Fate of the Subversive Science

Michael P. Cohen

There is one mention of Paul Shepard's work in all of my writing: "Paul Shepard has argued that the human skin is 'not a shell so much as a delicate interpenetration.'"[1] Yet everything lies behind that brief mention, because his phrase has been a key to my own inquiry about ecological thinking and consciousness I refer to "Introduction: Ecology and Man—A Viewpoint," in *The Subversive Science: Essays Toward an Ecology of Man*. The essay is not one of Paul Shepard's major works. Yet what I have to say here focuses on the boundary between the human and the world, and consequent problems he identified in that important essay.

As he explained in 1969, the idea of ecology is particularly subject to oversimplification. Especially: "It is not a discipline. There is no body of thought and technique which frames an ecology of man. It must be therefore a scope or a way of seeing." The value of ecological thinking, he understood, was in its perspective: "Man is the world, and his ecology is the nature of that inness."[2]

In the quarter of a century that has followed, we have seen considerable investigation of this issue of perspective, but we have not seen a solution to the problem Shepard so accurately described. "We are hidden from ourselves by habits of perception," he wrote.[3] And so we are. Our bodies know more, perhaps, than our words can say, because the structure of our discourse inscribes a distance from ourselves and constructs a shell around us.

Language itself encouraged us to perceive individually, Shepard argued, where any self or plant or animal became only "an isolated sack, a thing, a contained self." Our verbs and nouns divided our world and divided our selves, destroying engagement and connection, he said. Language, according to modern linguistics, creates meaning from differences. Yet what mattered about ecological thinking was the possibility of "vision across boundaries," in a "landscape as a creative, harmonious being where relationships of things are as real as the things."[4]

In this essay, Shepard pursued relentlessly and attacked the illusion of the centrality of modern man to his world—in part because "science as it is now taught does not promote a respect for nature." Shepard's idea of ecological thinking promised not answers, but humility, as framed by an unbounded set of questions. As he said, "The ideological status of ecology is that of a resistance movement."[5] The value of the Rachel Carsons and the Aldo Leopolds was, for him, in their "subversive" roles, as they challenged the public and private rights of citizens, scientists, and managers. Ecological thinking functioned as a critique, a critical method.

When he concluded "There is only one ecology," he spoke of something like the Tao, unnameable and unbounded, because neither human nor subhuman; it was uncontrollable by any school, theory, agency, or project.[6]

I am interested in the relevance of Paul Shepard's thinking on this subject to "nature writing," whose admirers (1) believe there is a something called nature, and (2) believe that humans can speak for it. Even if one accepts the first premise, the second does not follow from it, unless we ignore all modern thinking. Most recently, some have proposed a theory and practice of "ecocriticism," which seems to be attempting to say that (1) there is some essential or primary order of nature called "ecology" which we can know, and (2) with "ecology" in hand, we can speak for those texts which attempt to speak for nature.

Because I read Shepard early, I became skeptical of the dialogue humans are capable of having with the world, finding no reason to believe that nature speaks our language, or that we speak its language. Had we been having a monologue? I asked.[7] Being a student of literature and nature, I was led later to post-structural literary theory as it converged on the history of science. Shepard provided a good solid ground for this inquiry. As I considered the condition of modern monologue, explored in the 1960s by Michel Foucault, problems, as in Shepard, began not with the world's body, but with our own. Science's monologue was like that of medicine and reason, Foucault wrote:

> In the serene world of mental illness, modern man no longer communicates with the madman: on one hand, the man of reason delegates the physician to madness, thereby authorizing a relation only through the abstract universality of disease; on the other, the man of madness communicates with society only by the intermediary of an equally abstract reason which is order, physical and moral constraint, the anonymous pressure of the group, the requirements of conformity. As for a common language, there is no such thing; or rather, there is no such thing any longer; the constitution of madness as a mental illness, at the end of the eighteenth century, affords the evidence of a broken dialogue, posits the separation as already effected, and thrusts into oblivion all those stammered, imperfect words without fixed syntax in which the exchange between madness and reason was made. The language of psychiatry, which is a monologue of reason about madness, has been established only on the basis of such silence.
>
> I have not tried to write the history of that language, but rather the archaeology of that silence. [para. 2][8]

Why did this sound so familiar? Perhaps because it could be every kind of dialogue: the pattern is now universally familiar, which was surely the point. Transformed into terms more applicable to Shepard's subject, here is Foucault's issue. A key to transformation:

> madness = wildness
> mental illness = a different order
> (or) disease = disorder
> madman / man of madness = the wild
> physician / psychiatrist = resource manager

I am aware that such a transformation of terms resolves no philosophical issue, but it demonstrates the social construction of the problem Shepard perceived.

> In the serene world of (a different order), modern man no longer communicates with the (wild): on one hand, the man of reason delegates the (resource manager) to (wildness), thereby authorizing a relation only through the abstract universality of (disorder); on the other, the (wild) communicates with society only by the intermediary of an equally abstract reason which is order, physical and moral constraint, the anonymous pressure of the group, the requirements of conformity. As for a common language, there is no such thing; or rather, there is no such thing any longer; the constitution of (wildness) as a (different order), at the end of the eighteenth century, affords the evidence of a broken dialogue, posits the separation as already effected, and thrusts into oblivion all those stammered, imperfect words without fixed syntax in which the exchange between (wildness) and reason was made. The language of (the resource manager), which is a monologue of reason about (wildness), has been established only on the basis of such silence.
>
> I have not tried to write the history of that language, but rather the archaeology of that silence.

My transformed version of Foucault's first paragraph suggests the following about discourse and wildness. The age of reason substituted human monologue for a broken dialogue with nature. This is the cultural meaning of modern resource sciences, as the end of dialogue. They do not touch the subject of our inquiry—which is relations—but segregate it from us, and turn it to an object of reasoned discourse.

For Foucault, unlike Shepard, our current desire to recover our relations to the wild is belated and may be futile. As Foucault's principal American expositors point out: "Foucault [in 1969] seems to have thought that there was 'something' like pure madness which all these different cultural forms were groping after and covering up — a view he later abandons."[9] In the last two decades, scholars seem to have decided that there are no such things as pure madness, or pure wildness, or pure ecology. There are only academic discourses.

And Foucault was speaking of discourse. Arnold Davidson, one of his explicators, says that "truth is to be understood as a system of ordered procedures for the production, regulation, distribution, circulation, and operation of statements"—a set of discursive practices.[10] Which explains what Foucault identified as immature sciences—"too tinged with empirical thought, too exposed to the vagaries of chance or imagery, to age old traditions and external events, for it to be supposed that their history could be anything other than irregular."[11]

What Shepard welcomed Foucault disparaged, for surely the discourse of 1960s ecology, with its climaxes and its etymological households, with its webs of life, with its diversity and complexity, with its organic wholeness and harmony, has been exposed to the vagaries of chance and imagery. (See Daniel Botkin, *Discordant Harmonies*, for a critique of this discourse.)

Shepard had high hopes for the discourse of ecology in the late 1960s, which is why he and McKinley entitled the collection of "Essays Toward an Ecology of Man" *The Subversive Science*, borrowing the idea from Paul Sears, who explained the term in *Bioscience* (1964):

> My choice of title is not facetious. I wish to explore a question of growing concern. Is ecology a phase of science of limited interest and utility? Or, if taken seriously as an instrument for the long-run welfare of mankind would it endanger the assumptions and practices accepted by modern societies, whatever their doctrinal commitments?[12]

Shepard, of course, was a pioneer in exploring Sears' suggestion that ecology was a revolutionary tool. Our problem, he thought, "hidden from ourselves by habits of perception," would be remedied as we began to use "ecological thinking" which "requires a kind of vision across boundaries. The epidermis of the skin is ecologically like a pond surface or a forest soil, not a shell so much as a delicate interpenetration."[13] After nearly a quarter of a century, this promise has not been met by professional ecologists. There is no vision across boundaries, no delicate interpenetration. They describe edges, gaps, discontinuities. And environmental historians have come to no consensus about the "subversive benefits" of ecology. Consider the views of Donald Worster whose history of ecological science, *Nature's Economy: The Roots of Ecology* (1977), is the standard text. Most recently, he has expressed his concern that the persistent enthusiasm for ecology, represented on T-shirts, or in movements such as "Deep Ecology," suggests a hope that the science can offer "a pathway to a kind of moral enlightenment." Most of the political slogans of the "ecology movement" come from the language of Eugene P. Odum's *The Fundamentals of Ecology* (1953).

However, in studying the actual changes in science, rather than the slogans which have come out of it, Worster has seen a diversity of conflicting theories of ecological change appearing recently. And he concludes, in part, that "Ecology today, no more than in the past, can not be assumed to be all-knowing or all-wise or eternally true."[14]

Younger historians are even less optimistic. In a recent "Roundtable on Environmental History" in the *Journal of American History*, Richard White, new western historian, has critiqued Donald Worster's historical method—a method which has a kinship with the method of many so-called ecocritics. White asserts that any history which posits deductively a base of natural history, a structure of productive relations, and a superstructure of culture and ideology in order to generate cautionary tales is suspect. It is suspect because it reveals under archeological analysis a remarkable similarity to the structure of Marxist ideology in its system of ordered procedures for the production, regulation, distribution, circulation, and operation of statements.[15] Should historians make ecology the base of natural history, and the foundation for method? As White comments disparagingly, "ecology is, in fact, only an academic discipline."[16] He has decided that ecology is, as Sears had denied, "a phase of science of limited interest and utility."

On the other hand, younger environmental historians have also adopted Shepard's idea of ecological thinking. William Cronon, in "The Uses of Environmental History," (1993), lists the following four "deepest articles of faith," which he shares with his colleagues. "1. All human history has a natural context." "2. Neither nature nor culture is static." "3. All environmental knowledge is culturally constructed and historically contingent—including our own." "4. Historical wisdom usually comes in the form of parables, not political recommendations or certainties."[17]

Finally, Shepard, like modern environmental historians, doubted that the historical object being studied would have an "essence." Ecology was a tool: "We must use it to confront the great philosophical problems of man—transience, meaning, and limitation—without fear," Shepard said.[18] We must investigate with care different human techniques, as responses to local objectives.

Yet there is this final and decisive difference between Shepard and post-structural thinkers. For someone like Foucault, there is no thing we can call nature, no process we can call wildness, no place we can call wilderness, no order we can call ecology. Yet Shepard reminds us that nature, process, wildness, wilderness, and ecology exist, because the world exists in itself, whatever humans think, and our ideas can exist as perennial ways of understanding our relations with the world we are in. Where did scientific ecology go astray? Perhaps at its birth. What do we discover in the birth of a scientific ecology if not the following:

> ...small techniques of notation, of registration, of constituting files, of arranging facts in columns and tables that are so familiar to us now. . . . One should look into these procedures of writing and registration, one should look into the mechanisms of examination, into the formation of the mechanisms of discipline, and of a new type of power over bodies. Is this the birth of the sciences of man? It is probably to be found in these "ignoble" archives, where the modern play of coercion over bodies, gesture and behavior has its beginnings.[19]

Well, one might say, Foucault was speaking of training and disciplining human bodies, but, with the aid of a simple key, the science of the human body becomes the science of the world's body. Our new and immature science of ecology has grown by using the methods and rules of discourse of traditional science. As medicine explored the power/knowledge relations of human bodies as subjects and objects, now ecology is rooted in domination, subjugation, the relationship of forces—in a word, power.

What then of Foucault's enigmatic second paragraph, and the archaeology of silence? This has been the task to which Paul Shepard's work has called us. Though no society has ever adapted perfectly to its environment, though our thinking cannot be based on the findings of scientific ecologists or facile "Nature writers," yet it is the silence that calls us, requiring what he defined as ecological thinking.

Notes

1. Michael P. Cohen, *The Pathless Way: John Muir and American Wilderness* (Madison: University of Wisconsin Press, 1984), 153.
2. Paul Shepard, "Introduction: Ecology and Man—A Viewpoint," in Paul Shepard and Daniel McKinley, eds. *The Subversive Science: Essays Toward an Ecology of Man* (Boston: Houghton Mifflin, 1969), 1.
3. Ibid., 2.
4. Ibid., 3, 4, 5.
5. Ibid., 6, 9.
6. Ibid., 10.
7. Beware: in a monologue, the appearance of rational ordered analysis hides the failure of discourse to reach its object or communicate with another subject.
8. Michel Foucault, *Civilization: A History of Insanity in the Age of Reason* (New York: Vintage, 1988), xxi.
9. Hubert L. Dreyfus, and Paul Rabinow, *Michel Foucault: Beyond Structuralism and Hermeneutics* (Chicago: University of Chicago, 1982), 4.
10. Arnold Davidson, in David Couzens Hoy, ed., *Foucault: A Critical Reader* (New York: Basil Blackwell, 1986), 221.
11. Michel Foucault, *The Order of Things* (New York: Random House, 1970), ix.
12. Shepard, *Subversive Science*, v.
13. Shepard, "Introduction," 2.
14. Donald Worster, "The Ecology of Order and Chaos," *Environmental Historv Review* 14.1-2 (Spring/Summer, 1990):1-18.
15. Richard White, "Environmental History, Ecology, and Meaning." *Journal of American History* 76.4 (1990):1111-1116.
16. Ibid., 1115.
17. William Cronon, "The Uses of Environmental History," *Environmental History Review* 17.3 (Fall 1993):1-22.
18. Shepard, "Introduction," 10.
19. Michel Foucault, *Discipline and Punish: The Birth of the Prison* (New York: Vintage, 1979), 190–91.

Recovering Green Identity: Paul Shepard's Archeology of Self

Robert Hood

INTRODUCTION

AS RICHARD RORTY POINTS OUT, "Interesting philosophy is rarely an examination of the pros and cons of a thesis. Usually it is, implicitly or explicitly, a contest between an entrenched vocabulary which has become a nuisance and a half-formed new vocabulary which vaguely promises new things."[1] Paul Shepard's implicit appeal to "green" character virtues provides a new vocabulary for framing the relationship between human identity and wilderness in terms which are sensitive to the diversity of cultural perceptions and narratives. I claim that implicit in Shepard's work is an appeal to environmental virtues and vices, and to the notion of environmentally good persons. Shepard's archeology of Western identities, showing the importance of animals and wild places to the development of self-consciousness, destabilizes the margins between "civilization" and "wilderness," in part because it appeals to sentiments already held by members of contemporary American culture. Because his boundary explorations serve to destabilize the codes of Western civilization concerning the relations between humans and wilderness, Shepard's work can be seen as a strategy of resistance against the homogenization and domestication of Euroculture.

ONE WAY OF FRAMING ENVIRONMENTAL ETHICS

Kenneth Sayre argues that current environmental ethics are generally of three sorts.[2] They (1) apply a preferred standard ethical theory, such as Kant's deontology, to environmental problems, or they (2) extend the class of morally considerable beings to nonhumans, or they (3) adopt a theory of moral pluralism which involves mixing and matching the first two approaches.[3] In all three cases, environmental ethicists have responded with theories. Sayer argues that this theoretical, "inferential approach contributes little to the pressing need which environmental philosophers have been attempting to address in recent decades—the need for a set of normative values actually in place within industrial society that will help preserve the environment from human destruction."[4] The chief reason that theories are not effective is that theories as such are not useful in persuading and motivating people who are not already so moved. It is unlikely that rules alone can change people's decision-making process (as opposed to the justification of such a process) and make people care about the consequences of their successes or failures to follow rules. As Sayre points out, the theorist's involvement in environmental ethics presumably was motivated in the first place by an awareness that environmental norms of some sort ought to be in place: "It is not clear how the cause of environmental ethics has been advanced by a process of theoretical reasoning leading to a conclusion of which the theorist was unaware previously."[5]

One reason why inferential approaches do not persuade mass publics is a stylistic or rhetorical one. In another context Martha Nussbaum notes that almost without exception Anglo-American philosophical prose is "hygienically pallid," and that certainly a mistake is made when "one takes a method and style that have proven fruitful for the investigation and description of certain truths—say those of natural science—and applies them without further reflection or argument to a very different sphere of human life that may have a different geography and demand a different sort of precision, a different norm of rationality," such as environmental ethics.[6] A questionable underlying assumption of inferential approaches is that careful scholarship should be disinterested because emotions cloud rational judgment. However, experiencing certain emotions is a constituent part of accepting a certain belief or beliefs. Our reaction to someone who in total seriousness says they "love the environment" but feels no distress at the sight of trees being cut down to enlarge a parking lot is that they do not know what they are talking about. At least, our reaction will be to question that individual's sincerity, either of the individual's belief-claims or the individual's denial of emotion, because such a belief entails feeling distress under certain conditions. Nussbaum's claim is that different criteria of rationality are involved when one is coming to hold a view, or coming to accept a view, than when one is justifying a view already held. The criteria encouraged among professional environmental ethicists (such as academic disinterestedness and objectivity), which are important to the project of justifying environmental beliefs, are largely irrelevant to the process of coming to hold such beliefs oneself or encouraging others to do so.

Another reason why abstract moral principles fail to motivate sentiment is that the process of identifying rules itself presupposes an internalized awareness of moral norms.[7] Whether or not I intend to follow a rule depends not so much on neutral criteria but on whether or not I judge a rule to be coherent with the kind of person I want to be. Thus, the willingness to follow certain rules presupposes certain notions of human identity and certain concomitant virtues. Rule-based approaches may not be entirely fruitless. They at least allow professional academic philosophers to get tenure and thus institutionalize environmental ethics as a "serious area of academic inquiry." But they serve mainly to formalize and justify, rather than change, moral perception.

Eugene Hargrove argues that another problem with rules is that inferential approaches limit the range of ethical decision making, turning moral decisions into mutually exclusive alternatives. Hargrove's claim is that principles which are universalizable and prescriptive in ethics could constitute a body of rules over which the moral agent as moral agent has no choice. To violate them would then mean that the moral agent is no longer engaging in moral activity. While it is no doubt necessary and appropriate to have some rules which specify such limits, they, nevertheless, remain completely incompatible with the activity of decision making, since they limit rather than guide the decision-making process.[8] Such rules limit ethical decision-making processes because, rather than encouraging the development of more sophisticated interpretations of moral perception, rules tend to exclude. In attempting to simplify moral decision-making, those who invoke an inferential approach to environmental ethics actually end it and merely justify antecedently held beliefs.

A final reason against the inferential approach to environmental ethics is that such an approach does not seem to be working. New ways of thinking and acting are needed because inferential approaches are not successful by themselves in persuading those not antecedently committed to environmental reform to alter their behavior. Inferential approaches are not externally successful in motivating masses of people to change their behavior. Nor are they internally successful in resolving the interminable debates among contemporary environmental philosophers themselves over such issues as intrinsic value or moral pluralism. To summarize: the majority of environmental ethicists (the dominant culture) is engaged in a project of justifying, rather than creating, environmental concern. Rhetorical strategies based upon an inferential approach should be supplemented with others that create and nurture environmental concern because (1) discussions of rules are not suited, stylistically and rhetorically, to developing sentiment; (2) decisions to follow rules presuppose rather than create relevant character traits; (3) rules limit the range of ethical decision-making; and (4) inferential approaches are not succeeding in creating widespread concern for the environment or in resolving disagreements among environmental philosophers.

AN ALTERNATIVE WAY OF FRAMING ENVIRONMENTAL ETHICS

The chief alternative way of framing ethical reasoning involves an appeal to moral and intellectual virtues.[9] In recent years there has been a reappraisal of virtue ethics and a growing recognition of their importance for environmental ethics.[10] Unlike ethicists who formulate and apply rules of moral behavior, virtue ethicists focus on paradigmatic persons who exemplify environmental virtues as a means of developing moral perception and character in others. By identifying environmental virtues and by discussing paradigmatic persons such as Henry David Thoreau and Aldo Leopold, virtue ethicists try to make clear just who is an environmentally good person. Unlike ethicists who hold that the path to environmentally good behavior entails applying and following rules, virtue ethicists hold that the goal is to identify and emulate environmentally virtuous actions and persons. Implicit in Shepard's archeology of the deep past and his call for "A Post-Historic Primitivism" is an appeal to certain virtues. His work also suggests solutions to some problems with environmental virtue ethics.

One possible problem with environmental virtue ethics concerns how to distinguish environmentally virtuous actions from those which are not. Geoffrey Frasz notes that environmental ethics raise "epistemological problems" when "the new tradition does not share the same standards of what is good with the previous one."[11] Because virtue ethics are based upon right actions, a dilemma can arise concerning just what constitutes a right action when there are several incommensurate traditions present, as is the case in contemporary American society. Shepard's answer to the question of how to adjudicate between conflicting traditions is informed by an anthropological perspective. For Shepard, a right action is one that is ecologically sustainable over long periods of time. Hence, the environmentally unsound practices of twentieth-century peoples, such as poor land management practices and sprawling urbanization, are not virtuous because they are not sustainable.

Unfortunately, arguments such as Shepard's for adopting ecologically responsible practices are often seen as "utopian" or "Romantic." One way to counter such arguments is to emphasize the virtue ethics position implicit in Shepard's thought. If it can be shown that not only do environmentalists value certain virtues and try to emulate certain environmental heros, but that the same virtues and character traits have been valued for thousands of years, then responsible environmental policy would be harder to characterize as "Romantic" or "utopian."[12] On my view, the chief importance of Shepard's work is that his recovery of the deep past provides empirical evidence that similar character virtues are common to ecologically responsible paleolithic peoples and to contemporary environmentalists.

Another traditional problem with virtue ethics, which led Modern philosophers to reject them in favor of inferential (e.g., deontological, consequential, or utilitarian) approaches, is that traditional virtue theories are based upon the notion of the consistency and basic universality of human nature, a position that is difficult to maintain in a pluralistic society. However, recent reevaluations of virtue ethics, as in Alasdair MacIntyre's *After Virtue*, can be seen as a bridge between traditional virtue ethics that worked best in a closed and relatively stable culture and the need for a revised ethic that applies to postmodern cultures congealing around plurality and difference. MacIntyre avoids the critiques of traditional virtue ethics by showing it is possible to have a notion of virtue that is not predicated upon an inherent, ahistoric notion of human nature. MacIntyre recontextualizes human identity in terms of a character in a narrative drama and as a character involved in certain narrative practices.

For MacIntyre, humans are, in their actions and practice, as well as in their fictions, storytelling animals.[13] Thus, the fundamental context of human moral agency for MacIntyre is a narrative, rather than biological, one. MacIntyre holds that it is impossible to make sense of human moral identity outside the context of a historical narrative, and argues that "narrative history of a certain kind turns out to be the basic and essential genre for the characterization of human actions."[14] By engaging in such a narrative history MacIntyre is able to redescribe traditional accounts of the virtues, replacing biological conceptions of human nature with a concept of human identity contextualized within a narrative tradition. MacIntyre's work is just one example of a growing recognition among moral philosophers that a person's identity is a constructed character not unlike those in novels, plays and other storied communication. By inhabiting different sorts of stories and by institutionalizing them "quite literally," as Anthony Paul Kerby notes, by "rewording ourselves" we become different sorts of moral beings.[15] Rather than a biological fixity, the self exists as a site of narrative contextuality.

If human moral identity is contextualized within a narrative context as a scripted character, then there can be no necessary, single description of personhood, no final vocabulary to end conversations regarding what constitutes a good person. The turn toward narrative and the recognition that as our cultural and narrative traditions change through time, so will we, suggests a way of recontextualizing moral discourse in postmodern cultures of difference. Namely, as our narratives change, so will our moral and cultural identity. From this frame, language is not something we have, but something we are. That is, the relation between discourse and self is internal—the horizons of one's discourse are largely the horizons of one's self. The postmodern "self," rather than being a perdurable entity, is instead constituted through cultural and linguistic contexts: "the self is a social and linguistic construct, a nexus of meaning rather than an unchanging entity."[16]

Two possible objections to this postmodern way of characterizing the self and moral identity are (1) it may seem to suggest that individuals can adopt any narrative they wish and (2) any narrative tradition is as good as any other. However, these objections are specious for several reasons. First, language is not a private phenomenon. If we want to be understood by others we must make the same kinds of noises they make, or at least show them how to relate their familiar noises to our unfamiliar ones. Moreover, as Shepard recognizes in his own work, language, culture and the environment are internally related and cannot be separated. While it is possible to inhabit a wide variety of stories, some are better suited to the environment than are others. Indeed, this is one area on which both postmodernists and Shepard share common ground: both are trying to displace the hegemonic discourse in contemporary industrial cultures concerning Technology, History, and "Progress." When Shepard critiques History as "the notion of a fixed essence, an inner state that persists in spite of the contradictions of appearance," he is engaged in a projects similar to Jacques Derrida's deconstruction of Hegel's mythic conception of History and Michel Foucault's genealogy of the "totalizing discourses in current bureaucratic institutionalizations."[17] In short, the strength of Shepard's archeology of human moral identity is that he provides a rich array of empirical data about environmentally good way of living across cultures and across time. I contend that Shepard's recovery of the deep past is of more than purely academic interest. Rather, it holds moral lessons about what constitute "green virtues" and environmentally good persons.

In addition, it provides examples that individuals can emulate. The chief reason that the notion of green virtues should be pursued is that they seem to be the ones which already appeal to many environmentalists. Professional environmental ethics have had relatively little impact on average environmentalists, who continue to look to the writings of and about environmental "heros" such as Ed Abbey, Aldo Leopold, John Muir, and Henry David Thoreau for support. Environmentalists want model people they can emulate, not model cases they can study, and it is just these examples of model people, across cultures and across time, that Shepard's archeology provides.

SHEPARD'S SACRED PAW AND THE RECOVERY OF GREEN IDENTITY

Within a postmodern interpretive frame of virtue ethics, changing the dominant narrative whether by reinterpreting it, or by displacing it is one way of changing moral identity. Environmental virtue ethics, as seen from this frame, should not formulate rules but rather should try to reinterpret or displace existing narratives. Shepard's archeology of the Western self and moral identity in *The Sacred Paw* does just this; namely, he demonstrates how a green sensibility is already part of existing narratives.[18] By a green sensibility I mean a series of virtues which help people lead environmentally good lives. The following pages summarize Shepard's claims that: (1) interaction with non-human others and its depiction in narrative traditions helped form and legitimate certain notions of personal identity; (2) changes in the depictions of non-humans in narrative traditions parallel changes in Western identity; and (3) the linguistic traces of such interactions within oral cultures remain within the discursive horizons of Western literate peoples. Then I discuss several character virtues that are implicit in Shepard's archeology and that together form a green sensibility. In short, I argue that Shepard's reinterpretation of existing narrative traditions (e.g., the traces of the bear on Western literate consciousness) can expand the cultural conversation and encourage several green character virtues that may help us care for marginalized non-human others.

In *The Sacred Paw* Shepard demonstrates there has been a close association between humans and bears since the earliest records of human existence. The record of the bear's role in the birth of human consciousness is recorded in archeological sites, but also in ordinary language. In order to emphasize the integral relation of the bear to Western language and identity, I quote the following passage at length:

> The Teutonic stem word *ber* is the forerunner of *bar*, the German for "bear," and such terms as "barrow," "burgh," "barley," "beer," "berth," "bereave," "berate," "beruse" (or lullaby), "berg," "burgher," "bier," "bairn,"; such proper names as the cities of Bern, Berlin, and Bergen; and personal names like Bernard, Bertha, Gilbert, Herbert, Robert, and liter-

> ally, the Scandinavian name *Biorn* or *Bjorn*. Other extensions from the stem syllable produce *pare* and *pero*, or "parent" and "father," which are in turn related to generate and procreate. A series of connected words, *berjusjis, beruseis, bruseis,* tie the Teutonic stem word to the Latin oasis or Ursus as well as to *orso* and *oso*, "bear" in Italian and Spanish, through emus to *arctos*, "bear" in Greek. In this way "arctic," as well as personal names such as Arthur, may be seen as bearish in their origins, the latter being the half-legendary king of medieval England, who was said not to be dead but only sleeping [i.e., hibernating] on the isle of Avalon. *Ber* is also cognate with *her* and *ver*. From *fer* comes ferrum, the Latin for "iron," as well as "feral," along with "ferry," "fertile," and "ferocious." From *verfortis* is derived the Icelandic world eofor for "man," or eorn which, as bjorn, means "formed," and is also "bearn" in Danish. As for the Greek *arktos*, a cognate word is *arkhos* or "chief," with its scores of related words such as "archaic," "archbishop," and "archeology," all referring to "overarching" primary sources and powers.... The syllable from which comes "bear" also produces "basket," "amphora," and, perhaps most important of all, "metaphor," meaning literally to "carry over."[19]

What should be apparent from this linguistic evidence is that the bear is central to Western consciousness. Indeed, the verb "to bear," whose family of meanings share the sense of "to carry," is an example of what linguists call primary words, those words whose meaning is irreducible to any simpler terms. But, granted that the bear has a central role in being human, what meaning has this association had in the past, and what meaning does it have for contemporary peoples?

Shepard argues that "The bear is the prototype wild animal, a kind of model of man in the wilderness, whose home is the mountain or 'berg,' where the bear people live, to whom the bear returns in hibernation and death."[20] There have generally been three metaphorics associated with the bear. The oldest is the myth of the Bear Mother. "The Bear Mother tale itself reveals that bears are kinsmen: true forebears. It tells of men descended from the Bear Mother, a woman who married (and became) a bear and founded a race of immortal beings ourselves and that men are part human and part divinity, since the Bear Prince she married was a shape-shifting, all-hearing immortal who knew and controlled the cosmic power of conducting the sacred hunt."[21] The myth has three settings: the sky, Earth's surface, and the underworld. The Great Bears in the sky, Ursa Major and Ursa Minor, occupy a significant portion of the heavens and provide basic navigational guides to place and travel. The Earth's surface is the setting of the meetings between humans and bears in the myth, and also the setting of the daily tutelage the bear provided. The underworld is the realm of renewal: female bears give birth during hibernation and emerge from their den bearing young. These three mythic realms can be thought of as metaphors for different territories of the self—rational consciousness which plans or navigates (sky), the everyday upkeep and maintenance mode of being human (earth), and the chthonian realm of subconscious renewal (underworld). As Shepard notes, "renewal like interior has multiple meanings in which the bear remains a central figure. Our own interiors are also caves, inaccessible to the rational ego with its need for conscious light."[22]

The Bear Mother myth emphasizes a multivocal narrative of self, and portrays successful movements between them. As Shepard argues, although we are culture-dwelling creatures,

> much of the unconscious life of the individual is rooted in interaction with otherness that goes beyond our own kind, interacting with it very early in personal growth, not as an alternative to human socialization, but as an adjunct to it.... Identity formation grows from the subjective separation of self from not-self, living from nonliving, human from nonhuman, and proceeds in speech to employ plant and animal taxonomy as a means of conceptual thought and as model of relatedness."[23]

This multivocal notion of personal identity is due in part to the dynamics of oral cultures. Walter Ong points out that because self evaluation requires a partial demolition of situational, operational thinking and an isolation of the self, oral cultures do not have the highly individualistic notion of human identity characteristic of contemporary Western literate cultures. For primary oral people there is no self separate from the conversational community, including the land and its people.[24] As Duerr argues, we "should turn wild so as not to surrender to our own wildness, but rather to acquire in that way a consciousness of ourselves as tamed cultural beings."[25]

One way of looking at the transformations of the bear myths is that this multivocal notion of personal identity is domesticated and limited over time. An important part of the process of domestication involves the shift from oral to literate consciousness. Major changes in the bear myths occur in the contexts of the increasingly literate peoples of ancient Greece and medieval Europe. For example, the bear is imaged in contradictory ways in the bestiaries, medieval texts designed to communicate moral lessons to common people through animal personae. On one hand, the bear is portrayed as existing beyond civilized life, with a massive and uncontrolled appetite, and without regard for rules or law. The bestiaries also contain a contradictory image of bears as the symbol of maternal care and affection. The connection between the two made by the Church is that the bear symbolized the Christian in an unformed state. "Every bear possessed ...[a] devilish character; but the Church could redeem him."[26] The bear "came ironically to stand for Christianity itself. After all, the Church provided the means for spiritual shaping; it offered guidance and presented a model for the novitiate. The Church was, moreover, responsible for rescuing its members from moral and spiritual death, for reforming and regenerating heathens."[27] By identifying paganism with the wildness of the bear and Christianity with the domestic elements of the bear, the Church legitimated a different notion of human identity. Namely, one which valorized domestication. The people who participated in the Bear Mother myth lived in a culture which valorized otherness; in contrast, medieval society fenced out wildness. Those who experienced their otherness were labeled berserkers, that is "bear-like" and were considered crazy. The denigration of wildness, of otherness in general, parallel the rise of the effects of literacy in creating an increasingly homogeneous mass consciousness.[28]

The domestication of humans and bears reaches its apex in the persona of the teddy bear. Shepard notes that "from pagan sexual bear to female Christian bear to neuter stuffed bear the animal had traveled the entire gender circuit."[29] The teddy bear was created when Teddy Roosevelt refused to shoot a cub or a sow on a bear hunt in 1902. The teddy bear is a simulacra, a term used by Jean Baudrillard to point out the negative consequences of the tendency to believe that the abstractions of text-based cultures and the secondary orality of the media are more real than lived experiences. As Baudrillard puts it, "The very definition of the real has become: that of which it is possible to have an equivalent reproduction.... The real is not only what can be reproduced, but that which is always already reproduced. The hyperreal...which is entirely in simulation."[30] Confusing the wildness of bears with its simulation in teddy bears involves a kind of self-deception. As Tom Birch argues, it is a deception motivated by a desire for power: "the whole point, purpose, and meaning of imperial power, and its most basic legitimation, is to give humans control over otherness."[31] The notion of human identity legitimated by teddy bears is one based on the self-deception that bourgeois market simulations of wildness can satisfy the human need for true wildness. Rather than a opportunity to experience the other, to cross the fence between wilderness and civilization, teddy bears are yet another extension of the dominant culture's domesticating power.

Implicit in Shepard's narrative about the bear from the deep past to the current time is a series of green virtues that were once common but that have become increasingly marginal. However, it is simplistic to say that paleolithic peoples "valued wilderness" both because it is likely that the conceptual bifurcation between "civilization" and "wilderness," so common to industrial cultures, does not apply to archaic peoples, and because it does not explain why they valued wild nature. This is the importance of Shepard's work: the virtues contained in the bear narrative explain why wild nature has value. Moreover, this narrative is important because it shows that the virtues valued by contemporary environmentalists are not new, but are shared with paleolithic peoples. The following discussion of green virtues, by no means exhaustive, is culled from Shepard's work on paleolithic cultures and is intended as a first step to recover a green identity and to nurture environmentally good persons.

GREEN CHARACTER VIRTUES

One of the more important fictions of the Modern world is a clear separation between civilization and wilderness. Implicit in this view is that there is a clear separation between the domesticated and wild aspects of the self as well, a view apparent in the Christian opposition between a perdurable, redeemable soul and a corrupted, corporal body. The insights of Freud and depth psychology notwithstanding, there is a reluctance to recognize that "Identity formation grows from the subjective separation of self from not-self, living from non-living, human from non-human, and proceeds in speech to employ plant an animal taxonomy as a means of conceptual thought and as a model of relatedness."[32]

Perhaps the most important virtue of paleolithic peoples is the virtue of liminality, the willingness to engage in psychic boundary explorations and encounter with the Other. The virtue of liminality is predicated upon the recognition of symbiotic relations between the human and the nonhuman and the recognition that it is not desirable for the human animal to exist as only wild or only domesticated. Paleolithic peoples recognized that "one could become 'tame' only if before one had been 'wild', that one could only live in the true sense of the world if one had proved one's willingness to die. In order to be able to live within the order, in other words, in order to be consciously tame or domesticated, one had to have lived in the wilderness. One could know what inside meant only if one had once been outside."[33]

However, as Oelschlaeger notes, "Our prevailing definitions of 'wildness' and 'wilderness' preclude recognition of nature as a spontaneous and naturally organized system in which all parts are harmoniously related; in consequence, humankind has believed itself compelled to impose order on nature."[34] Archaic cultures recognize the virtue of boundary-crossing, of moving between realms of order and disorder to comprehend each more fully. They realize we "should turn wild so as not to surrender to our own wildness, but rather to acquire in that way a consciousness of ourselves as tamed, as cultural beings."[35] One example of this boundary-crossing between the human and non-human world can be seen in the orientation of archaic cultures around the bear. The liminal rituals of archaic cultures such as vision quests are socially sanctioned ways to encourage the process of getting free from oneself and recognizing relation to the nonself.

Another green character virtue is the virtue of going native, which involves encountering wild nature and experiencing it for its own sake. Shepard notes that the "crucial question of the modern world is 'How are we to become native to this land?'"[36] Indeed, in contrast to contemporary cultures where Sartrean alienation and existential sweats are common products of urban industrial life, a strength of paleolithic cultures is the sense of belonging to the land that occurs as a result of daily, physical contact with wild nature. Perhaps one of the better examples of a person coming to realize the virtue of going native is Henry David Thoreau. In 1846 Thoreau journeyed to Ktaadn, the highest peak in Maine. It was a journey which had pivotal importance:

> Perhaps I most fully realized that this was primeval, untamed and forever untameable Nature, or whatever else men call it.... Talk of mysteries! Think of our life in nature, daily to be shown matter, to come in contact with it, rocks, trees, wind on our cheeks! the solid earth! the actual world!

the common sense! Contact! Contact! Who are we? Where are we?[37]

Thoreau's experience on Ktaadn focused his comprehension of the interrelation between humankind and nature. More importantly, as Oelschlaeger notes, "Ktaadn rekindles for Thoreau a primal or paleolithic coming-to-consciousness of humankind's naked rootedness in and absolute dependence upon nature."[38] It would be a mistake to interpret Thoreau's experience from a modern frame and dismiss it as a debased sensual experience. Although it was sensual, it provided important information about his relation to the cosmos—information which could not have been gained from disinterested contemplation. Indeed, the sensual education provided by going native is its virtue—for it creates the connections that motivate ethical concern for wilderness. As Thoreau noted, "In Wildness lies the preservation of the world"—that is, in the sensual experience of going native lies the source of the psychological connections needed to care about nature.

GREEN INTELLECTUAL VIRTUES

Whereas green character virtues involve practices which lead to habitual action in accord with environmental goods, green intellectual virtues are the theoretical and practical skills necessary to reach green ends. One intellectual virtue of Shepard's Post-Historic Primitivism is the virtue of orality. There are important structural differences between oral and literate cultures. Oral communication is an inherently dynamic and primarily a communal event. In a culture with writing, communication is also possible through books and letters—inherently individual pursuits, unless one is reading to a group. But in an oral culture communication necessarily occurs in groups. Not only does oral communication encourage group solidarity, because ideas can be shared only in groups, it also encourages solidarity through a shared awareness of sacred presence or power. As Walter Ong notes, "A hunter can see a buffalo, smell, taste and touch a buffalo when the buffalo is completely inert, even dead, but if he hears a buffalo, he had better watch out: something is going on."[39] When the primacy of oral discourse is forgotten, words cease to be events and become placemarkers, that is, abstract references to things. Making this distinction between the words-as-events of oral cultures and words-as-placemarkers of written cultures is important, because when typographic cultures such as our own think of words not as events, and instead appropriate them as mere placemarkers for things, things too are appropriated and become mere resources.[40]

Unlike textual discourse, which emphasizes analysis and innovation, oral thought is aggregative and emphasizes preserving the story-cycle. Oral thought emphasizes narrative instead of argument. As Ong points out, the use of epithets and cliches is an important aspect of oral traditions. "Traditional expressions in oral cultures must not be dismantled: it has been hard work getting them together over the generations, and there is nowhere outside the mind to store them.... Without a writing system, breaking up thought—that is, analysis—is a high risk procedure."[41] The oral cultures of the paleolithic are also aggregative in a second sense, because ideas can only be communicated in conversation; unlike text-based or chirographic cultures, in oral cultures discussion of ideas always occurs in a group. The close-knit social structure of paleolithic peoples is not simply a consequence of belief but a reflection of the simple fact that oral communication is intrinsically social.

Whereas in reading it is possible to reread in order to clarify, in oral discourse there is nothing to refer back to except memory, because oral utterances vanish as soon as they are uttered. The redundant or copious aspect of oral discourse keeps both speaker and hearer on track. In order to be certain that the message is communicated, it is repeated: messages are repeated to be certain that they are comprehended. Also, because gaps and pauses in oral delivery are distracting, speakers use repetition to avoid hesitation.

Oral discourse is conservative in the sense that it preserves tradition. Because conceptualized knowledge which is not repeated aloud soon vanishes, oral cultures encourage repetition of traditional stories and discourage experimentation. "Knowledge is hard to come by and precarious, and society regards highly those wise old men and women who specialize in conserving it, who know and can tell the stories of the days of old."[42]

Oral cultures remain in contact with tactile, lived experience. Ong points out, for example, that oral cultures have nothing as neutral as a list. "Oral cultures know few statistics or facts divorced from human or quasi-human activity."[43] Rather, in hunter-gatherer bands, experience is organized around totems. Shepard notes that "plant and animal totems are not simply objects of reverence: they serve as codes for formulating and relating the conceptual systems of nature and of culture."[44]

Discourse in oral cultures is agonistically toned; that is, knowledge is situated within a context of struggle. Proverbs and riddles provoke strong responses—the goal of agonistic discourse is to emotively engage others. "When all verbal communication must be by direct word of mouth, involved in the give-and-take dynamics of sound, interpersonal relations are kept high—both attractions and, even more, antagonisms."[45] In short, if knowledge depends for its existence on being told, then it must be interesting to tell. Hence, the moral proscriptions of oral cultures often occur within stories which involve clear distinctions between good and evil, using the characters of villains and heroes to convey the message. The discursive dynamics of oral cultures legitimate a different notion of being human. Ong points out that because self-evaluation requires a partial demolition of situational, operational thinking and an isolation of the self, oral cultures do not have the highly individualistic notion of human identity characteristic of contemporary Western literate cultures. For paleolithic peoples who comprehend the virtue of orality there is no self separate from the conversational community, including the land and its people.[46] For example, one man when asked about himself merely placed his hand over his heart and responded, "What can I say about my own heart? How can I talk about my character? Ask others; they can tell you about me.

I myself can't say anything."[47] To summarize: Shepard's "voice of life" is important because oral discourse is inherently communal and encourages human interrelations—what Gary Snyder calls the "primacy of together-hearing."[48] In addition, the dynamics of oral discourse encourage a sense of community with nature.

Also important is the intellectual virtue of living according to a hunting-gathering economy. The hunting-gathering economy of paleolithic cultures has the advantage of minimizing impact on the land; paleolithic peoples lived in small, mobile bands rather than in densely populated, sedentary urban environments. Hunger-gatherers stay in one place for relatively short periods of time and do not impact an area beyond its ability to recover. Perhaps one of the more important consequences of the hunting-gathering economies is the hunt itself. The confrontation with death in hunting nurtures what the Greeks called sophrosyne or the "skill of mortality." As Shepard notes, the hunt "signifies the gift of life, the obtaining and preparation of which ritualizes the encounter of life and death, in which the human kinship with animals is faced in its ambiguity, and the quest of all elusive things is experience as the hunt's most emphatic metaphor."[49] When every member of the band is intimately involved with killing and dressing game it is difficult to achieve the same level of abstraction about nature common to urban mass publics in supermarkets.

RECOVERING GREEN IDENTITY BY REFRAMING ENVIRONMENTAL ETHICS

Shepard's work provides a context against which it is possible to recognize the extent of our domesticated identity. When teddy bear stories such as Winnie the Pooh are read against the Bear Mother myths, the difference is obvious. The contrast helps us recognize the extent to which character virtues that emphasize sensuality and the body have been domesticated. Shepard's archeology of oral consciousness helps us comprehend the identity legitimated by the majority of environmental ethics articles. Namely, Shepard's work helps us see that rule-based environmental ethics is a colonizing and totalizing discourse which actually continues to legitimate the character virtues of civilization. Rule-based approaches do not help us step out of who we are and nurture the character virtues and intellectual virtues needed to care for wildness. Yet in a time of ecological crisis this epistemological shape-shifting is imperative—for what is at issue is persuading others, not justifying existing apathy. As Duerr notes, at such times, "and whether this is more a psychological argument than an epistemological one we need not argue at the moment, we have even have to sacrifice the ability to represent such experiences in a generally intelligent way in return for achieving the ability to experience them."[50] Shepard's narrative approach to Western identity gives us an alternative in green character and intellectual virtues and a range of other identities we might inhabit on the path to environmentally good lives. In a time of ecological crisis, projects such as Shepard's which use alternative frames to motivate concern may be more important than developing rules to justify existing beliefs.

Notes

1. Richard Rorty, *Contingency, Irony, and Solidarity* (Cambridge: Cambridge University Press, 1989), 9.
2. Kenneth M. Sayre, "An Alternative View of Environmental Ethics," *Environmental Ethics* 13 (1991): 195-213.
3. For examples of applications of standard ethical theories to environmental problems, see Michael Zimmerman, "Toward a Heideggerean Ethos for Radical Environmentalism," *Environmental Ethics* 5 (1983): 99-131; Susan Armstrong-Buck, "Whitehead's Metaphysical System as a Foundation for Environmental Ethics," *Environmental Ethics* 8 (1986): 241-259; William Chaloupka, "John Dewey's Social Aesthetics as a Precedent for Environmental Thought," *Environmental Ethics* 9 (1987): 115-145; Christina Hoff, "Kant's Invidious Humanism," *Environmental Ethics* 5 (1983): 63-70; and Brian Norton *Why Preserve Natural Variety?* (Princeton: Princeton University Press, 1987). For an example of attempts to extend the class of morally considerable beings, see Peter Singer, *Animal Liberation: A New Ethics for our Treatment of Animals* (New York: Avon Books, 1975); for an example of moral pluralism see Christopher D. Stone, "Moral Pluralism and the Course of Environmental Ethics," *Environmental Ethics* 10 (1988): 139-154.
4. Sayre, "An Alternative View of Environmental Ethics," 195.
5. Ibid., 202.
6. Martha Nussbaum, *Love's Knowledge: Essays on Philosophy and Literature* (New York: Oxford University Press, 1990), 19-20.
7. For further discussion of rules, see Ludwig Wittgenstein, *Philosophical Investigations,* trans. G.E.M. Anscombe (New York: Macmillan and Co., 1953).
8. Eugene C. Hargrove, "The Role of Rules in Ethical Decision Making," *Inquiry* 28 (March 1985): 25.
9. For my purposes, an environmental virtue is a necessary but not sufficient trait of intellect or character, the possession of which allows a person to lead an environmentally good life. Green intellectual virtues are those theoretical and practical skills needed to achieve a good life; green character virtues are promoted by practices which lead to habitual action in accordance with environmental goods.
10. See Eugene Hargrove, *Foundations of Environmental Ethics* (New York: Prentice Hall, 1989), Thomas E. Hill, Jr., "Ideals of Human Excellence and

Preserving Natural Environments," *Environmental Ethics* 5 (1983), 211-224, and Geoffrey B. Frasz, "Environmental Virtue Ethics: A New Direction for Environmental Ethics," *Environmental Ethics* 15 (1993), 259-274.

11. Frasz, "Environmental Virtue Ethics," 263.
12. Such a connection is still important even though the dominant cultures from neolithic times to the present generally have not valued nature.
13. Alasdair MacIntyre, *After Virtue: A Study in Moral Theory*, 2d ed. (Notre Dame: Notre Dame University Press, 1984), 216. My emphasis.
14. Ibid., 208. MacIntyre's emphasis upon human history as "the basic and essential drama" fails to take into account the textual and cultural limits of this reading. Compare Paul Shepard, "A Post-Historic Primitivism" in *The Wilderness Condition: Essays on Environment and Civilization*, Max Oelschlaeger, ed. (San Francisco: Sierra Club Books, 1992) and Hans Peter Duerr, *Dreamtime: Concerning the Boundary Between Wilderness and Civilization*, trans. Felicitas Goodman (New York: Basil Blackwell, 1987), esp. 116-124.
15. Anthony Paul Kerby, *Narrative and the Self* (Bloomington: Indiana University Press, 1991), 64.
16. Ibid., 34.
17. Shepard, "Post-Historic Primitivism," 43-44.
18. Paul Shepard and Gary Sanders, *The Sacred Paw: The Bear in Nature, Myth, and Literature* (New York: Viking Press, 1985).
19. Shepard, *Sacred Paw*, xiv.
20. Ibid.
21. Ibid, 59-60.
22. Ibid., 69.
23. Paul Shepard, *Nature and Madness* (San Francisco Sierra Club Books, 1985).
24. Walter J. Ong, *Orality and Literacy: The Technologizing of the World* (London: Routledge, 1990), 55.
25. Duerr, *Dreamtime*, 125.
26. Shepard, *Sacred Paw*, 132.
27. Ibid., 139
28. See Marshall McLuhan, *The Gutenberg Galaxy: The Making of Typographic Man* (Toronto: University of Toronto Press, 1962).
29. Shepard, *Sacred Paw*, 165.
30. Jean Baudrillard, "The Precession of Simulacra," in *Art After Modernism*, ed. Brian Wallis and Marcia Tucker (Boston: David R. Godine, Inc., 1988), 253.
31. Thomas H. Birch, "The Incarceration of Wilderness: Wilderness Areas as

Prisons," *Environmental Ethics* 12 (1990), 18.

32. Shepard, *Nature and Madness*, 125.
33. Duerr, 43.
34. Max Oelschlaeger, *The Idea of Wilderness: From Prehistory to the Age of Ecology* (New Haven: Yale University Press, 1991), 8.
35. Duerr, 125.
36. Shepard, *Post-Historic Primitivism,"* 52.
37. Henry David Thoreau, *The Maine Woods* (New York: Penguin Books, 1988), 93-95.
38. Oelschlaeger, *Idea of Wilderness*, 149.
39. Ong, *Orality and Literacy*, 32.
40. Ong notes that "Such 'things' are not so readily associated with magic, for they are not actions, but are in a radical sense dead...." *Orality and Literacy*, 33.
41. Ibid., 39
42. Ibid., 41
43. Ibid., 43
44. Paul Shepard, *The Tender Carnivore and Sacred Game*, (New York: Charles Scribner's and Sons, 1973), 203.
45. Ong, *Orality and Literacy*, 45.
46. Ibid., 55.
47. Ibid., 46.
48. Snyder quoted in Shepard, "Post-Historic Primitivism," 87.
49. Shepard, *Post-Historic Primitivism*, 82.
50. Duerr, *Dreamtime*, 130.

Deep Ecology, Depth Psychology, and the Call of the Ecological Unconscious

Daniel Sloan Brown

It was twenty years ago (1974) that I received a catalog from an unconventional college in southern California. As I turned through the pages, I took special note of a photograph of a man alongside a short monologue about the environmental crisis and its relationship to human consciousness. I was immediately drawn to this man and the ideas that he was able to share at this unique school. He would eventually come to embody for me the intellectual necessity of moving beyond the orthodox boundaries of compartmentalized knowledge. I had just finished my first year of undergraduate work at a midwestern university and was being groomed by the Psychology department to be a die-hard behaviorist, with all the naive assumptions of prediction and control. At this particular school, behaviorism was touted as the environmental psychology of the times and was given the credit of having moved off the unreliable couch of the psychoanalyst and into the "real" world of stimulus and response. I was able to maintain my enthusiasm for the possibilities of behavior modification for a second year at that midwestern school, but the alternative college in California beckoned me with the added resource of sociology, anthropology and authentic environmental sciences. And then there was that professor in the catalog and his references to consciousness and its relationship to the archaic ways of our past. I received the 1975 edition of the Pitzer College catalog and there he was again, more poetic references to the nature of consciousness and something about the ontological necessity of the nonhuman in our childhood and well-being. I transferred to Pitzer in the autumn of 1975 and the most tangible expectation that I had was to meet this provocative man and explore what he had to say.

I lived an academic double life at Pitzer. While preparing myself to become a dissector of brains (a neurologically enhanced behaviorist), I began to take those classes on the side that had caught my attention in the catalog. I was accompanied by my roommate, Peter, who had just finished traveling for over a year in far away lands and who had been most affected by the indigenous peoples he had encountered. He was looking for a way to understand the numinous magic he had felt; I was reexamining my assumptions about what made us human and conscious. In the few precious years allotted us in this undergraduate, time-critical phase of development, my friend and I took every course that this man offered. The eloquence of his speech and thought never failed to expand our sense of who we were, where we were coming from, and what may have been missing from our lives. Our self-made ritual was to go to the local Thai restaurant after each and every class and try to contain our excitement over a dish of Thai fried-noodles. We would gleefully bandy about terms we had learned on that day and add them to our ever expanding taxonomic system of understanding human nature. Anyone sitting in hearing range of our conversations would have heard phrases like "peripheral male," "ontogeny recapitulates phylogeny," "time critical receptivity to archetypal potentials," and "the simultaneous spiralling of separateness and relatedness with the natural world." The rest of my studies paled in comparison to this world that had opened up to me; by graduation, this powerful educational experience left what would be a significant and permanent impression on my psyche. My friend and I have remained in communication since, and it is rare that we do not evoke the name of Paul Shepard and the vision that he shared with us in those days of our extended adolescent yearnings. (We also recall with great fondness the flock of Red Burma Jungle Fowl that he proudly shared with us on our all too rare visits to his canyon home.) I have been in academia most of my life in one form or another. At each successive stage of my academic rites of passage, I always seem to find one man that inspires me and moves me on to higher levels of meaning and values. Paul Shepard was that man in my college years. The influence of his teachings has only begun.

> Perhaps we do not need new religious, economic, technological, ideological, esthetic, or philosophical revolutions. We may not need to start at the top and uproot political systems, turn life-ways on their heads, emulate hunters and gatherers or naturalists, or try to live lives of austere privation or tribal organization. The civilized ways inconsistent with human maturity will themselves wither in a world where children move normally through their ontogeny....This means that we have not lost, and cannot lose, the genuine impulse. It awaits only authentic expression. The task is not to start by recapturing the theme of a reconciliation with the earth in all of its metaphysical subtlety, but with something much more direct and simple that will yield its own healing metaphysics.
>
> —Paul Shepard, *Nature and Madness*, 1982

The guiding theme of this paper is that the ecological crisis of our time is a kind of collective psychopathology that has evolved for thousands of years and is currently escalating at a feverish rate. The transfigurations that Western consciousness in particular has undergone in the last 12,000 years, have ushered it and its co-evolving dark side, "the unconscious," to our modern era where it is now provoking the most alarming feedback from the nonhuman world that it has ever witnessed. If we are to assume that there is something pathologically wrong with the modern mind that takes root in historical antecedents both personally and collectively, how can we as psychotherapists and other modern day shamans best intervene in the situation? Psychology has traditionally polarized itself into two generalized schools of thought that take different approaches to the resolution of people problems. One camp says that problems are primarily behavioral and cognitive and require new learning to displace old dysfunctional habits that have lost their adaptive value. The other camp has taken a more interior and biographical route to the heart of our problems by emphasizing the prerequisite of accessing and reexperiencing elements of the past that are somehow linked to present day problems. Depth psychology is the generic term applied to the latter camp and that is how I shall refer to it in this paper. The differences between these two camps are essentially ones of emphasis: relearning or reconditioning behavior and attitudes usually involves coming to terms with dysfunctional learning from the past, while working affectively and consciously with past events and traumas clears the way for new choices and positive action. The metapsychological question is always: which emphasis is most effective in ameliorating human suffering?

It goes without saying (for many) that the ecological crisis requires a massive overhaul of our dominant social paradigms and of our child raising practices. Redirecting our educational resources to the relearning of sustainable living strategies in all of their myriad forms is essential. On a more subtle and perhaps far-reaching scale, we need to revision the learning environments of the youngest among us and model more of their education on what their developing minds are intrinsically groping for. But I must leave these tasks to the educators among us; my concern is with the possible contributions that depth psychology can make towards the transformation of our cultural madness and ecological naivete.

If much of our collective psychopathology and ecological crisis is defined in stark contrast to an archaic past that involved a higher standard of relatedness between the human and nonhuman, how are we to reclaim such wisdom on a mass scale? Is ecological consciousness something that we can only construct anew through political and social forces based on environmental and economic necessities? Or are there preexisting psychic forces that reside in our problem child, the unconscious, that await some kind of stimuli or activation? Theodore Roszak raises this issue in his book, *The Voice of the Earth*, and suggests that there is indeed an "ecological unconscious" that lies at the core of the psyche and that can be revived as a resource in our efforts to transform our ecological destiny. He is quick to emphasize the speculative nature of his hypothesis, but then, that is what psychological theories are best at.

I personally do not see this as a radical or even novel idea. Innuendos of the existence of such a layer to the unconscious have existed at least since Freud first put forth his thesis of the psychological compromises between nature and culture. Freud, of course, was more akin to a biologist than to an ecologist, and he based his vision of the mind on the mechanistic metaphors of his day: energy, pressure, hydraulics, and so on. To his way of thinking, the human mind was primarily patterned by the physiology that it depended on, but it was simultaneously infiltrated by whatever culture it found itself in. The demands, expectations, and prohibitions of culture tend to be irresistible by virtue of our enhanced survival as a member of a family, tribe or community; thus, does the anxiety begin. The crucible of the mind becomes the battleground for the forces of nature against the juggernaut of civilization. It was the great double bind from which, Freud concluded, there was no satisfactory resolution. To submit to civilization and its modifications of our human embodiment of natural forces is to surrender to a life of discontent, anxiety and melancholy, yet, to resist this civilized mandate is to risk estrangement, isolation and the very means for our survival.

The unconscious was originally seen as the refuse of this pervasive process: a seething cauldron of impulses, instincts and desires that were hidden from our awareness because of the threat it held against the construction of our civilization. The unconscious and the repressed were basically one and the same. Eventually Freud came to see that not all that is unconscious is repressed by the individual; there were aspects of the unconscious that were unconscious at birth. As he developed his theory to include the ego and the id, he asserted that there were unconscious, unrepressed aspects to both. It was perhaps his attempts at the interpretation of dreams that led Freud to the implications of an "archaic unconscious."

> Whence comes the necessity for these phantasies, and the material for them? There can be no doubt about the instinctual sources, but how is it to be explained that the same phantasies are always formed with the same content? I have an answer to this which I know will seem to you very daring. I believe that these primal phantasies...are a phylogenetic possession. In them the individual...stretches out...to the experiences of past ages. [They include] abbreviated repetitions of the evolution undergone by the whole human race through long drawn out periods and from prehistoric ages.
>
> —Sigmund Freud, *The Standard Edition of the Complete Psychological Works of Sigmund Freud,* 1953-1969

This was as far as Freud would go in the exploration of transpersonal and transcultural realms. It would eventually become a bone of contention between Freud and his heir apparent, Carl Gustav Jung.

Jung was the first follower of Freud to dramatically alter and expand on the notion of the unconscious. He asserted the existence of a "phylogenetic heritage" that consisted of Freudian instincts in conjunction with mental entities or images that he called the "archetypes." He believed archetypes to be inherited through some unspecified medium and that they reflected species-wide patterns of behavior, thinking and feeling. They were archaic to his way of thinking and had arisen from the same prehistoric cauldron that the equally ephemeral instincts had exuded from. In his own words:

> Man inherits these images from his ancestral past, a past that includes all of his human ancestors as well as his prehuman or animal ancestors; these

racial images are not inherited in the sense that a person consciously remembers or has images that his ancestors had. Rather, they are predispositions or potentialities for experiencing and responding to the world in the same ways that his ancestors did."

—C. G. Jung, *A Primer of Jungian Psychology,* 1973

Some archetypal forces were viewed as a nuisance that tended to interfere with the Jungian objective of "individuation," a term for the process of attaining the highest form of maturity, while others, such as the archetype of the Self, were essential to this lifelong process. However, to my knowledge, neither Jung nor Freud ever discussed the possibilities or ramifications of specifically accessing ancestral aspects of the unconscious in the pursuit of health and well-being. Yet, they spent their lives developing their respective archeologies of the mind and in the process, made countless observations of the natural and precivilized being peering through the veil of the personal unconscious.

It would take another favored student of Freud's who also would eventually leave the house of psychoanalysis in order to establish a new vision if human nature; but this vision would equate what is good in man to what is natural in man. Wilhelm Reich took Freud's original thesis of the tension between nature and culture and extended their arena of conflict to include the whole of man's psychosomatic being. He made tangible what Freud could only assert in abstract terms. Reich's analysis of character development and his discovery of organismic armoring in man added a somatic wrinkle in our understanding of our alienation from nature in its external forms (landscapes and ecosystems) as well as its internal forms (pulsations, sensations, emotions). Reich was not limited to observations of the somatic side of life. Unlike most analysts, he was drawn to the study of primitive people, such as the Trobriander tribe of New Guinea. It was investigations such as these that led Reich to a respect for the relatively unarmored and graceful lifestyle that hunting-gathering people seemed to enjoy. These were people spontaneous with their emotions, affirming of their sexuality and connected with their natural surroundings. Reich took great joy in asserting to the psychoanalytic community that there was no evidence of the oedipal complex or other psychoanalytic artifacts in non-patriarchal societies, and that Freud's pessimistic, entropic assessment of humankind was not necessarily the most logical one.

Reich was eventually expelled by the psychoanalytic society, but by then he was laying the groundwork for a new understanding of the human condition that we now refer to as bodywork and somatic psychology. His assertions were simple and profound: the separation and antagonism between the "I" and the "IT" of the Freudian doctrine, had found its literal counterpart in the tapestry of sinew and muscle. The ego driven voluntary nervous system had evolved in conjunction with a constrictive web of chronic muscular tension that Reich believed was the primary "modus operandi" of repressive forms of enculturation. (Reich was especially observant of the socio-political characters of his day and wrote extensively on the somatic basis of fascist and communist totalitarianisms.) Conflict and suffering were not to be blamed on the aggressive and sexual drives of the id crashing into the boundaries of the ego, but rather, as byproducts of an egoic process bent on subverting what was natural and organic to the demands of culture. For Reich, there was a resolution, a way out of the trap, and it depended on accessing and releasing our natural proclivities towards love and work. This is not the paper to go into any detail, but Reich's theories and therapy would evolve over the years to become increasingly inclusive of the natural world. He would eventually come to see our psychosomatic nature as embedded in a planet-wide web of bioenergetic and ecological interrelationships. Indeed, the many analogies and correspondences he made between organismic functions and planetary processes was a harbinger of the Gaia hypothesis that was yet to come. He was particularly concerned with the encroachment of desertification and believed that this is how the landscape reflected our own "collective emotional desert." Reich's appreciation of hunting-gathering life, his integral approach to the body/mind dichotomy, and his scientific and intuitive grasp of our interconnectedness with the natural and nonhuman environment, qualifies him as one of the first "ecopsychologists" and as someone whose insights can be most informative in our quest to cultivate ecological consciousness in our time.

> Most people seem to agree that we cannot and do not want to go back to the past; but the reason given is often wrong: that time has moved on and what was can never be again. The truth is that we cannot go back to what we never left. Our home is the earth, our time the Pleistocene Ice Ages. The past is the formula for our being.
>
> —Paul Shepard, *The Tender Carnivore and the Sacred Game,* 1973

Depth psychologies have at least two things in common: history and the unconscious. Each of the aforementioned theoreticians pushed the envelope of the unconscious to include increasingly wider dimensions of human experience, while simultaneously deepening our understanding of the presence of the past in our unfolding ontogeny. Perhaps the most epistemologically important development in depth psychology recently has arisen from the work of Stanislav Grof. Grof was trained initially as a Freudian psychoanalyst and eventually was drawn to the use of psychoactive substances as a way to enhance and improve analytic work. He worked primarily with LSD and observed thousands of sessions, first in Prague and then eventually at the National Institute of Mental Health in Maryland. What he explored both affirmed and surprised what his training had led him to expect. Early LSD sessions with a client would reveal much of what a Freudian would look for in Western culture: oedipal dynamics, difficulties with aggression and sexuality, issues with intimacy and self-esteem. What he did not expect was the periodic eruption of psychosomatic experiences that subjectively included images and feelings of being immersed in a womb-like state, of separation and fall from such a state, and of a kind of life and death struggle that resulted in a sense of liberation and rebirth.

This happened so consistently with the clients that were undergoing progressively deeper explorations in the sessions that Grof began to consider the archetypal significance of such a recurring pattern. Perhaps this was indicative of a universal layer of consciousness where Jungian and Reichian phenomena merged into a psychic and somatic unity. This also revived questions that the psychoanalyst, Otto Rank, had raised in the previous generation about the trauma of birth and how it relates to the successive ontogeny of the child. Freudians and neo-Freudians have mapped the first six years of life extensively and assert that all the seeds of pathology are sown in these years; yet, most of the presenting problems in Grof's work had corresponding elements in the rebirth experience. An example would be that clients complaining of depression and feeling stuck in their life usually reexperience the phase of birth that involves unrelenting contractions with no anticipation of cessation. Clients with anxiety and obsessive-compulsive problems reexperience the phase that involves moving through the birth canal with all of its associated elements of blood, feces, and falling. The significance of these perinatal matrices of experience is not only how they relate to subsequent stages of development, but also the doors of perception and even deeper levels of the unconscious that their reenactment seem to open up.

It was as if this rebirth experience opened a Pandora's box full of images and sounds, peoples and places and most importantly, for our purposes, it evoked empathy and identifications with the nonhuman world around us. Grof refers to this level of the unconscious as the transpersonal or Holotropic realm. It is here that many clients and explorers have had vivid encounters with other species or have themselves taken on the forms, sensations and cognitions of animals. More than a few have plunged into a predator and prey scenario, apparently taking on the perceptual and emotional capacities of one or the other. Others have reported flashbacks to arboreal and savannah environments, some have even been able to recount sights, sounds, and smells of places they have never been in this life, but that others could confirm were coherently and accurately represented. Last, but not least, many people who have never heard of a Gaia hypothesis return from their Holotropic journey and report having felt a kind of communion with a unitary biospheric consciousness. These are not just "new age" California trendies submitting these recollections; these are doctors, lawyers, Harvard professors and others who had no premeditated agenda for having a "mind-blowing" experience.

At this point many may be quick to trivialize these experiences as the manufactured fantasies of an unnatural, drug-induced sensorium. But Grof's work evolved to include a drugless catalyst currently known as Holotropic Breathwork. For almost 20 years he and his students have used a combination of intensive breathing, Reichian inspired bodywork, Jungian inspired mandala drawing, drumming and music to evoke these levels of experience and the results have been the same. This continuity of experience, I believe, lends credibility to the notion that there is something intrinsic to our unconscious life that "awaits...an authentic expression." Whether the authenticity of these expressions that I have referred to be measured in literal, metaphorical, or mythological terms, is not my concern. My focus is on how they effect and move the experiencer. Do these altered states of consciousness translate into an expanded state of environmental action and ecological responsibility? If we have indeed a means to evoke an archaic or "ecological unconscious," what good is it if it isn't integrated into an ecological consciousness?

> The complexity of human development is so great and the scope for error so broad that we might wonder how anyone gets through to full maturity. Some repair processes are built into its calendar. The most conspicuous is adolescent regression, an infantile aspect that may serve the individual in recuperative ways. Beyond that, psychotherapy (which occurs in some forms in all societies) is usually shaped around a reexperiencing that enables the patient to adjust to, if not mend, a psychodevelopmental flaw. The catalysts of mind, provided by the nurturers, are like the nutritional and hormonal factors of the neotenic body, except that they are required experiences. Like the timed events in a play...they may be poorly presented, muffed, or missed. The play goes on, more or less imperfectly. The person, so deprived, bears the consequences as unregenerate elements; of immaturity, glossed by repression and compensation, distorted by unconscious yearnings for overdue fulfillment and resentments against the nurturers.
>
> —Paul Shepard, *Nature and Madness*, 1982

In *Nature and Madness*, Paul Shepard illustrates the epigenetic spiral, the "to and fro of autonomy and symbiosis," a summation of the first twenty years of ontogeny and its earthly relations. He describes development as a spiraling upward between separation and differentiation on the one hand and the necessary complementation of reconnection and higher-order relatedness on the other. The dynamics of the autonomous phases are well-documented and even encouraged by Western psychologies; what is less understood and appreciated are the phases of relatedness that serve to provide us with a meaningful context for our life journey. I would like to suggest that at the root of the epigenetic spiral, underlying the maternal and earth matrices that Paul Shepard identifies as essential symbiotic phases for the unfolding of maturity, there exists the perinatal matrices. One of the main tenets of the spiral model is that each successive level tends to recapitulate developmental themes from earlier levels. How then does the recapitulation of the perinatal matrix fit into the genetic karma of human experience? Why does its remembering often bring to the foreground of awareness images and emotions about our earthly relations? Perhaps there is a simultaneous activation of all the matrices (perinatal, maternal, ecosocial and cosmological) in their internalized, psychological forms and, analogous to the eastern concept of Kundalini rising, the energy of the base matrices of experience illuminates the intimations and yearnings for the higher order matrices. In our culture these successive matrices have been impoverished in their contextualizing and life-affirming power, particularly the maternal and earth matrices that are so important to the infant and adolescent phases of ontogeny respectively. It is conceivable that what is unfulfilled in the psyche is what asserts itself with imagery and emotion in Holotropic Breathwork.

The perinatal, maternal and ecological themes that characterize the Western mind's experience of this work may be indicative of what the psyche is lacking in our time and what it requires in terms of parental and social capacities in order to move on to maturity. The perinatal matrix is the original rite of passage; all subsequent rites of passage also involve an ending to some level of identification and a rebirth into a larger, more inclusive identity. If this personal evolution is not met at the critical stages with a culture's capacity to imbue it with meaning and purpose, we are obliged as psychotherapists and healers to discover and promote whatever remedial actions our phylogenetic inheritance allows.

We have heard the phrase "ontogeny recapitulates phylogeny." If this is true, what are the possibilities of a species-wide return to an ecological matrix of meaning and value in the wake of the agricultural, Judeo-Christian, scientific, military-industrial epochs of the last dozen millennia? Have these stages of "civilizing" the human animal been our cultures one-sided attempt to promote the autonomous, bounded, masterful self in search of control over birth, death, nature and the feminine? It is difficult to argue with an historical record that has delivered humankind to the moon; at least until one acknowledges the price that our ontogeny and our "participation mystique" have had to pay. Perhaps this separation from the "anima mundi," the soul of the world, is in some evolutionary way a prerequisite for a collective calling forth for what we have lost. Perhaps we stand at the threshold of a new cosmological world order that will usher our future generations into a higher symbiotic phase of relatedness that we can hardly imagine. Whatever form this evolution may take, it is unlikely that we can cast off the moorings of our archaic past as we seek to navigate into the waters of an ecological world view.

The Psychopathology of the Human-Nature Relationship

Ralph Metzner

SEVERAL DIFFERENT DIAGNOSTIC METAPHORS or analogies have been proposed to explain the ecologically disastrous split, the pathological alienation, between human consciousness and the rest of the biosphere. None of these psychological diagnoses, incidentally, have been made by psychologists, who seem to have taken absolutely no interest in this question thus far. From one point of view these concepts are metaphors, analogies transferred from the realm of individual psychopathology to the level of society, or even to the level of the entire species and its relation to the nonhuman natural world. From another point of view they could be seen as diagnostic concepts applied in the realm of collective or mass psychology, on a par with Wilhelm Reich's *Mass Psychology of Fascism* or Lloyd deMause's psychoanalytic interpretations of (mostly modern) historic events.[1] In either case, the purpose of such diagnostic speculation or intuition is the same: we are trying to discern the nature of the psychological disturbance which appears to have Homo sapiens in its grip, in order to be able to apply whatever we may have learned in psychology about the treatment of such conditions to the amelioration of the present ecocatastrophe.

Before proceeding to the psycho-pathological metaphors, let me first briefly discuss the diagnostic analogies in terms of organic pathology. A number of people have proposed that in view of the excessive population growth of humans in many parts of the biosphere, the best analogy to describe the situation is in terms of a malignant tumor: tumors are made up of cells multiplying uncontrollably and destroying the surrounding tissue. Anthropologist Warren Hern has pointed out, "A schematic view of the growth of London from 1800 to 1955 looks like nothing so much as an expanding, invasive, matastatic, malignant tumor."[2] He proposes the term Homo ecophagus ("ecosystem-devourer") as the appropriate name for this pathological species. In view of the obvious malignancy of the large megacities, Theodore Roszak has written about "Gaia's city pox."[3] It is worth noting that all such concepts depend on the following underlying analogy: that Earth (Gaia) is an organism and humans and other individual organisms are the cells in this super-organism. Human populations are then seen as metastasizing cancer cells, spreading to more and more areas of the Earth's land surface.

Al Gore has taken this disease metaphor to task, after first erroneously attributing it to deep ecology. He writes, "The obvious problem with this metaphor is that it defines human beings as inherently and contagiously destructive, the deadly carriers of a plague upon the earth. And the internal logic of the metaphor points toward only one possible cure: eliminate people from the face of the earth."[4] But this is an incorrect reading of the disease analogy: there are several different approaches to the treatment of cancer or virus infections, including treatments involving changes in diet, life-style, attitude, self-concept and imagery, all of which have shown some successes and have obvious implications for the collective human situation. There is also the well-documented existence of so-called "spontaneous remission," in which cancerous cells stop multiplying and tissues revert to a healthy condition, or in which the organism's immune system reduces the numbers of the pathogenic microbes to a tolerable level. The normal inhibitory mechanisms that prevent excessive growth can become re-activated. Could this be analogized to conscious communities of individuals restraining their reproductive rate?

The best articulated formulation of the disease analogy is by James Lovelock, in his last book *Healing Gaia*. Being trained in medicine, Lovelock has long been suggesting that geophysiology should be the name of the science of the structures and functions of the Earth's ecosystems and energy cycles (hydrogen, oxygen, nitrogen, etc.). Planetary medicine then concerns the disturbed functioning of this great organism. Lovelock's preferred diagnosis is that Gaia is suffering from a parasitical infestation by the species homo sapiens, a disease he calls disseminated primatemia. He points out that parasite-host relationships can have four possible outcomes: one, the invading microorganisms are destroyed by the host's immune system; two, host and parasite settle down to a long war of attrition—the condition known as chronic infection; three, the parasite destroys the host and thereby eliminates its own life-support; and four, the parasitical relationship is transformed into one of mutualism or symbiosis. "The last (scenario), symbiosis, is obviously desirable. As intelligent microbes, we have the advantage of knowing the risks of failure and the lasting benefits of symbiosis. But will we achieve it?"[5] Lovelock states that there are several strong precedents in nature for this kind of symbiosis between life-forms of very different scale; but adds that there are inherent properties of humans that make it difficult for us to act sensibly and achieve symbiosis within Gaia. With this statement we are in the realm of human behavior and attitudes as being the crucial lever of global change.

Environmental thinkers or ecophilosophers were the first to point to the crucial role of distorted human attitudes, beliefs and values in the generation of the ecological crisis. The concept of anthropocentrism (or homocentrism) was offered as a philosophical diagnosis of the human species' ecological maladjustment and biocentrism (or ecocentrism) as the corrective. This idea has been put forward by a number of 20th century philosophers, ecologists and writers, including Aldo Leopold, Paul Sears, Aldous Huxley, Loren Eiseley, Rachel Carson, Lynn White, Jr.,

Robinson Jeffers, Paul Shepard, Gary Snyder, Edward Abbey and others; forebears of this kind of thinking have been traced to such 19th century thinkers as John Muir, Henry David Thoreau and Ralph Waldo Emerson, as well as to the ethical cosmology of Baruch Spinoza.[6] The anthropocentric critique has been articulated most cogently by the Norwegian philosopher Arne Naess (himself strongly influenced by Spinoza) and by those identified with the philosophical orientation known as deep ecology, founded by Naess. To use the term "anthropocentrism" as a critique of the human attitude to nature in the modernist worldview parallels the use of "ethnocentrism" to critique racial discrimination and "Eurocentrism" to critique the colonialist exploitation ideology of Western culture.

Although the role of anthropocentric attitudes in creating or aggravating the ecological imbalance of industrial civilization is unquestionable, there are reasons to question the use of that term, since it actually covers two distinct meanings. "Anthropocentric" literally means "human-centered," and some critics have pointed that the human, like every other species, necessarily looks at the world from its own point of view, seeking to maximize its own survival advantages. A non-anthropocentric viewpoint, in this criticism, is both impossible and unnatural. But whether our human-centered perspective can be transcended or not is surely an empirical question—a question about the possibility of psychological change. Even if we assume that a homocentric attitude is a basic fact of human nature, this still leaves open the possibility that humans can learn to transcend their inborn homocentrism. This is in fact what Arne Naess, Warwick Fox and others propose with their notion of extending identification to the natural world. Perhaps this is something humans can do and animals cannot. Or perhaps animals, or some animals, can also empathically escape their species-centered points of view and look at the world through human eyes. Certainly this seems to be the view of shamanistic cultures, which have much lore and myth about how in ancient times humans could understand the language of animals, and animals could understand humans, even to the point of marrying them some times, as in the story "The Woman Who Married a Bear".[7] Interspecies empathic identification may have been the norm in the paleolithic world, a capacity that has become atrophied through disuse in humans.

But the philosophical critique of anthropocentrism really is saying more than deploring the fixation in a human-centered perspective, just as ethnocentrism implies more than a perspective centered on one's own ethnic group. There is an implicit meaning of assumed superiority and right to dominate others. This has been referred to as human chauvinism or human imperialism (two metaphors from the political arena) as well as speciesism—on a parallel with prejudicial ideologies such as racism, sexism, classism and nationalism. In each of these forms of prejudice, one group of human beings assumes its superiority to other beings (human or non-human) and therefore arrogates to itself the right to dominate and exploit those judged inferior. But it is really human speciesism that we are talking about here, the human assumption of superiority to other species, what the ecologist David Ehrenfeld called "the arrogance of humanism" in his book of that title.[8] This leads to the surprising and depressing conclusion that humanism, that much prized core-idea of Western culture, has become a form of collective pathology involving discrimination, prejudice and domination, analogous to racism, sexism, classism and nationalism—even although the first Renaissance humanist philosophers probably had no such development in mind.

Because this is quite a distinct idea from the fixation on the human-centered perspective, I suggest we call this diagnostic metaphor *the humanist superiority complex*. We do not have space here to go into the interesting cultural historical question of how this superiority complex came to be such a pervasive feature of the Western worldview. The "dominion" imagery of the creation myth in Genesis has been identified by many historians, including Lynn White, Jr. and Arnold Toynbee, as one obvious source. Some Christian theologians have argued that "dominion" really meant "stewardship"; but for many, even the notion of stewardship is still based on assumed human superiority. Do we really know enough to be stewards? Hunter-gatherers are clearly not stewards, which is a term derived from animal husbandry ("sty-warden"). Crude and over-simplified interpretations of Darwinian evolutionary theory lent seeming support to the notion of man as the superior animal, more evolved and more complex than other life-forms, and therefore, it was assumed, with both the knowledge and right to tinker with nature and use it as we saw fi.[9] Current evolutionary biology has of course shattered all those assumptions, eliminating even this last rationalization for the humanist superiority complex.

"Superiority strivings" or "will to power" were a central element in the developmental theory of Freud's contemporary Alfred Adler. Adler believed that a conscious feelings of superiority are always a compensation for an unconscious inferiority complex; and such inferiority feelings tend to arise normally in childhood, as a result of prolonged dependency and immaturity. Jung and other psychodynamic psychologists accept the notion of conscious attitudes compensating for unconscious ones, even if they don't accept the centrality of the superiority-inferiority complex. If this analysis applies to the collective ideology of humanist arrogance, we would look for unconscious fears and inadequacy feelings towards the natural world, underlying the attitude of conquest and domination. With this speculation, we come close to some of Paul Shepard's ideas on the loss of paleolithic relatedness and developmental initiation rites.

The first person to clearly articulate a psychopathological metaphor for the destructive and exploitative behavior of the human species toward the natural world was Paul Shepard, in his book *Nature and Madness*. Drawing on the work of psychoanalytic developmental psychologists such as Erik Erikson and Harold Searles, Shepard brilliantly dissected the cultural pathology of Western Judeo-Christian civilization as a case of arrested development, or what he calls ontogenetic crippling. He traced the progressive distortion of normal developmental pathways, which could still be seen in surviving hunter-gathering societies, through four historical stages: the agricultural domestication, the desert fathers, the puritans, and the founders of mechanistic science. In his 1992 essay "A Post-

Historic Primitivism," Shepard summarizes his earlier work by saying that "incomplete, ontogeny runs into the dead end of immaturity and a miasma of pathological limbos."[10]

A particularly interesting feature of Shepard's analysis is his discussion of the interplay between neoteny, the extended period of immaturity and dependency of the human child, and the ontogenetic support provided by culture. "Neoteny, the many years of individual immaturity, depends on the hands of society to escape itself. This mitigation of our valuable retardation is in part episodic and social, a matching of the calendars of postnatal embryology by the inventions of caregivers. Occasions make the human adult"[11] In the case of a species with such marked neoteny as the human, the failure or disappearance of culturally provided developmental supports would have particularly devastating consequences. In his use of paleolithic hunter-gatherer societies as models of ecologically balanced lifestyles, including child-rearing, Shepard seems to be saying that with the advent of domestication, circa 12,000 years ago, civilized humanity began to lose or pervert the developmental practices that had functioned healthily for hundreds of thousands of years.

He sees two stages where ancient patterns of development may have become chronically incomplete: infant/caregiver relationships and adolescent transition rites. The distorting process "first began with a slight twist in the life of the child, with events that may only have marred his capacity for elderhood and judgment.... The history of Western man has been a progressive peeling back of the psyche, as if the earliest agriculture may have addressed itself to extenuation of adolescent concerns while the most modern era seeks to evoke in society at large some of the fixations of early natality."[12] Shepard argues that agriculture increased the distance between the growing child and the non-human or "wild" world of nature: "by aggravating the tensions of separation from the mother and at the same time spatially isolating the individual from the nonhumanized world, agriculture made it difficult for the developing person to approach the issues around which the crucial passages into fully mature adult life had been structured in the course of human existence."[13]

In Erikson's developmental model, adolescence is the time when the child is enmeshed in a conflict between "identity and identity diffusion."[14] The notion of a species-wide fixation at the stage of early adolescence fits with the kind of boisterous, arrogant pursuit of individual self-assertion that characterizes the consumerist, exploitative model of economic growth, where the short-term profits of entrepreneurs and corporate share-holders seems to be not only the dominant value, but the only value under consideration. It also fits with the aggressive and predatory militarism and emphasis on the values and ideals of male warrior cults that have characterized Western civilization since the Bronze Age. Adolescents who have difficulty negotiating the turmoil of this stage often become, as Erikson writes, "remarkably clannish, intolerant and cruel in their exclusion of others who are 'different,' in skin color or cultural background."[15] Erikson pointed out how totalitarian doctrines have a special appeal to youths looking for solid identity structures: "The tempestuous adolescence lived through in patriarchal and agrarian countries ...explains the fact that their young people find convincing and satisfactory identities in the simple totalitarian doctrines of race, class or nation."[16]

The rites of passage in traditional societies provided guiding structures for negotiating the transition from the family matrix to the larger society. The progressive deterioration and loss of adolescent rites of passage in the modern age is well-known. Robert Bly has pointed out how even the minimal father-to-son apprenticeship bonding that used to exist prior to the Industrial Revolution has since been lost. The only manhood transition rites involving elders we still have, such as they are, are the boot-camp and combat initiations afforded by the military. Beyond that, there is only the stunted futility of attempted peer-group initiation, whether in the pathetic form of college fraternity hazing, or in the casual violence of juvenile street gangs, where 12-year olds carry hand-guns to school to avenge imagined insults to their "home" band.

Besides the loss of adolescent initiation rites, Shepard points to the "unity pathology" that develops if the earliest stage of infant/caregiver bonding is disrupted or disturbed. This is the stage Erikson identifies as the stage where the child's developing sense of self is dealing with issues of "basic trust vs. mistrust." If this stage is not negotiated successfully we may have at best an attitude of chronic insecurity, and at worst the suspiciousness and violence-proneness of the paranoid psychotic. As Shepard says, "the social skills of the newborn and the mother's equally indigenous reciprocity create not only the primary social tie but the paradigm for existential attitudes."[17] Jean Liedloff's studies of mother-infant bonding among the Amazonian Indians and her continuum concept[18] support Shepard's assertion that studies of babies and parents in hunter-gatherer societies reveal that the intense early attachment leads not to prolonged dependency but to a better functioning nervous system.

Shepard summarizes his theory of ontogenetic crippling by stating that "Men [presumably he means 'Western industrialized humans'] may now be the possessors of the world's flimsiest identity structure—by paleolithic standards, childish adults."[19] One of the worst consequences of this collective pathology is "a readiness to strike back at a natural world that we dimly perceive as having failed us." Adults who have basic trust that the world of nature and society can provide for their needs, are not likely to be attracted to a worldview that demands a relentless struggle for competitive advantage. Government leaders and opinion makers in the United States are now in the habit of promoting "competitiveness" as the value or goal which the educational system should develop in the nation's children. We are suffering, Shepard says, from "an epidemic of the psychopathic mutilation of ontogeny."[20]

Shepard does not say much about the possible treatment for such a case of collective fixated development. Presumably the re-institution of initiation rituals for adolescents, carried out by respected elders, and a much greater sensitivity to the fragility of early infant bonding should be two key ingredients in any attempt to reverse this pathology. Similar proposals have been made by many people. "An ecologically harmonious sense of self and world is...the inherent possession of everyone; it is latent in the organism, in the interaction of the genome and early experience. The phases of such early experiences, or epigenesis, are the legacy of an evolutionary past in which human and nonhuman achieved a healthy rapport."[21]

A related psychopathological metaphor put forward by theologian-turned-geologian Thomas Berry is that the human species has become autistic in relationship to the natural world. He traces the origin of this autism to Descartes' invention of the mechanistic worldview: "Descartes...killed the Earth and all its living beings. For him the natural world was mechanism. There was no possibility of entering into a communion relationship. Western humans became autistic in relation to the surrounding world." Like autistic children, who do not seem to hear, or see, or feel their mother's presence, we have become blind to the psychic presence of the living planet and deaf to its voices and stories that nourished our ancestors in pre-industrial societies. "This situation can be remedied only by "a new mode of mutual presence between the human and the natural world."[22]

Psychiatrists still debate whether infantile autism is the same or different as childhood schizophrenia. The current version of the diagnostic manual of the American Psychiatric Association, the DSM-III-R, calls it a "pervasive developmental disorder," characterized by "qualitative impairment in reciprocal social interaction, ...qualitative impairment in verbal and nonverbal communication and in imaginative activity (such as role- playing, fantasy), ...and markedly restricted repertoire of activities and interests."[23] Stereotyped movements and behavior, restricted range of interests, obsessive routines, preoccupation with parts of objects, absence of imaginative play and lack of awareness of the feelings of others are all typical of autistic children. These characteristics can readily be observed in many adults of industrial culture, when compared to those brought up in vernacular societies.

The cause of infantile autism is not known; earlier views that it was caused by deficient mothering have given way to the general belief that it is a biochemical brain disorder. Some autistic children respond to vitamin B6 therapy; others to heroic and prolonged efforts by caregivers to dissolve the perceptual-affective barriers. Most are untreatable. It is clearly an extreme form of developmental deficit—and if this diagnosis of our cultural malaise is indeed correct, the prospects for humanity are not good.

A third analogy from psychopathology that offers considerable insight, in my view, is the model of addiction. We are a society whose scientists and experts have been describing for forty years, in horrifying and mind-numbing detail, the dimensions of global ecocatastrophe. Just think of some of the book-titles: *Silent Spring, The Population Bomb, The Limits to Growth, The Death of Nature, The End of Nature*. Our inability to stop our suicidal and eco-cidal behavior fits the clinical definition of addiction or compulsion: behavior that continues in spite of the individual knowing that it is destructive to family, work and social relationships. This metaphor of addiction or compulsion, on a vast scale, also parallels in many ways the teachings of the Asian spiritual traditions, especially Buddhism, which have suffering or dissatisfaction as an inevitable feature of all human consciousness and craving or desire at the root of suffering.

The person who has developed the addiction diagnosis most extensively is the deep ecologist and mountaineer Dolores LaChapelle in her book *Sacred Land, Sacred Sex*. In a chapter entitled "Addiction, Capitalism and the New World Ripoff," she analyzes the interrelationships between the pursuit of addictive substances, including gold, silver, sugar and narcotics, and the explosive spread of the capital accumulation industrial growth society all over the globe from the 16th century onwards to the present. "The entire development of capitalism consists in making a group of people addicted to some 'substance' and selling it to them. Capitalism 'worked' as long as we had an enormous source of cheap natural resources (primarily in the New World).... Continuing its history of 'addiction', capitalism is now relying more and more on addictive drugs to fuel its growth."[24] Several other authors have also pointed to the addictive nature of our relationship to fossil fuels, another major engine of unrestrained industrial growth and ecological destruction. More generally, one can see the spread of consumerism and the obsession with industrial-economic growth as signs of an addictive society.

Descendants of the pirate bands of the 17th and 18th centuries, modern transnational corporations are operating essentially as high-tech bandits, plundering the biosphere, always and only looking for the highest profit rate for their investors. As corporations, they are legal fictions, capital accumulation machines, with not even human interests, values or ethics to restrain them, much less any concern for the intrinsic value of non-human beings. One could say they are operating from a money-centered ideology. Buckminster Fuller called the multinationals the modern incarnations of mythic supra-human giants. Nevertheless, since they are giants created by humans, there does exist the somewhat hopeful possibility of humans dismantling or transforming these corporate giants—presumably this the self-appointed task of the "corporate responsibility" movement.

Yet another analogy is the notion that we as a species are suffering from a kind of collective amnesia. We have forgotten something our ancestors once knew and practiced—certain attitudes and kinds of perception, an ability to empathize and identify with non-human life, respect for the mysterious and humility in relationship to the infinite complexities of the natural world. It may be that at several crucial turning points in the history of human consciousness we chose a particular line of development and thereby forgot and neglected something—with fateful consequences. I take this to be one aspect of Paul Shepard's view: we forgot the adolescent initiations, we forgot the humility and relatedness of the paleolithic hunter-gatherer, we forgot the perceptual sensitivity to the ceaseless cycles of energy transformations occurring in the natural world. Paul Devereux and his collaborators, in their book *Earthmind* (Devereux 1989), write, "for a long time now, we have been unable to remember our former closeness with the Earth. Due to this amnesia, the ecological problems now thrust upon us have come as a shock.... We notice the emergence of an amnesia that is really a double forgetting, wherein a culture forgets, and then forgets that it has forgotten how to live in harmony with the planet."[25]

An elaboration of the amnesia metaphor would be to consider the possibility of a traumatic amnesia. We know from studies of the effects of child abuse and rape, of reactions to combat in wartime, of accidents and natural disasters, that where the person experiencing the trauma is in a completely helpless and out of control position, that the memory of the experience can be completely lost—even although physical effects on the body and symptoms such as nightmares and panic attacks may remain. Such buried memories can often be recovered with hypnosis or psychedelic psychotherapy. If this paradigm applies to humanity's amnesia for knowledge of interdependent relatedness that seems to have been normal and natural for paleolithic culture, and to a degree always maintained by certain individuals and groups up until the present time, then the question that presents itself is: was there some event that in a terrifying way threatened this sense of relatedness and harmony?

One person who proposed and defended this view with verve and brilliance was the psychoanalyst Immanuel Velikovsky, who in his book *Mankind in Amnesia* argued that planetary near-collisions in historic times caused massive and violent earth changes, leading to almost total amnesia and permanent fear and insecurity among humans.[26] Even if we don't accept his theory of cosmic catastrophe, there are plenty of candidates for extremely violent natural cataclysms, such as volcanic and seismic events, with widespread loss of life and forced migrations, during the last four to five thousand years, as well as earlier. For example, in the 15th century B.C.E. a volcanic eruption on the island of Thera in the Eastern Mediterranean, accompanied by earthquakes and tidal waves and of awesome destructiveness, obliterated the highly advanced ancient Minoan culture—and may have provided the historical prototype for the legend of sunken Atlantis. Other possible events causing traumatic amnesia may have been periods of prolonged rain and freezing, prolonged drought and aridity, sudden weather changes or invasions by marauding warrior-bands. In medieval Europe it is not difficult to imagine the traumatic effect of Christianity's prolonged onslaught on the pagan nature-cults, as well as the Black Death, which wiped out one third of the population in the 14th century.

The amnesia analogy in general is more hopeful than some of the others, since it is clearly much easier to remember something that we once knew, than it is to develop an entirely new adaptation. We can also see that the indigenous peoples of the Fourth World, whether in North and South America, Southeast Asia or Australia, have been trying for some time to help us remember certain vital attitudes and values that they have preserved and maintained in their own ways of life.

Theodore Roszak, in his book *The Voice of the Earth* has argued that ecology and psychology need each other, and that "repression of the ecological unconscious is the deepest root of the collusive madness in industrial society; open access to the ecological unconscious is the path to sanity."[27] Roszak points out that Jung's idea of the "collective unconscious" originally included pre-human animal and biological archetypes, but later came to concentrate primarily on panhuman religious symbols. He proposes that we take the original meaning and call it the "ecological unconscious," as "the living record of cosmic evolution." This may turn out to be a terminology that has a wide appeal, although I personally prefer Robert Lifton's idea of a "species self." Calling some image or understanding "unconscious," or even more, reifying it as "the unconscious," may function to keep it unconscious. After all, we are trying to foster ecological consciousness, or "ecological conscience," to use Aldo Leopold's term.

Roszak wants to rehabilitate or revision the Freudian id: instead of the predatory, lecherous beast of the founder of psychoanalysis, he thinks it could be seen as the repository of ancient ecological wisdom. "The id is the Earth's ally in the preservation of the biosphere ...(and) Gaia gains access to us through the door of the id." But this idea will not do, I believe, what he wants it to do. While it is true that our Western modern child-rearing practices effectively stifle any innate ecological sensibility the child may have, it is also true that in traditional societies ecological knowledge and respect for nature are passed on from parents and elders to children: they do not emerge from adolescence without such training. That is one of the reasons why the disruption of traditional cultures has been so environmentally devastating. "Open access to the ecological unconscious," whatever that may mean, is not going to be sufficient for a path to sanity, unless supplemented by a recovery of ancient traditions of initiation and ritual celebration, and a strong dose of ecological literacy.

In contrast to the Freudian and post-Freudian view of the centrality of repression in the creation of "the unconscious," there has been in recent years a revival of interest in the concept of dissociation. Dissociative disorders such as "post-traumatic stress disorder" (PTSD) and "multiple personality disorder" (MPD) are being diagnosed much more frequently, though it is not known whether this is due to an increase in the actual occurrence of such disorders or to improved recognition of conditions previously misunderstood. Dissociation is a normal and natural cognitive function, the opposite of association. Dissociation plays a role in hypnotic and other forms of trance, when we progressively disconnect perception of the external world, in order to attend to interior images, memories and impressions. Even the simple act of focussing or concentrating attention clearly involves some degree of dissociation.

In the Freudian view, psychic material (thoughts, images, feelings, etc.) which is in the repressed unconscious (also called id) is disorganized, primitive, and childish, functioning according to the "pleasure principle"; whereas the conscious mind (ego) functions according to the "reality principle" and is capable of adjusting or adapting to the "demands" of reality in a rational, organized manner. The dissociationist view, as originally put forward by Freud's contemporary Pierre Janet, and in the neo-dissociationist theory of Ernest Hilgard and others, is that dissociation involves a "vertical" separation of strands of consciousness that may be equally well organized, rational, and in touch with reality.[28] For example, the mental and emotional components of a painful experience may be dissociated, so that we remember what we saw and thought, but not what we (appropriately) felt; or, vice-versa, a certain stimulus may trigger a feeling state of panic, but the cognitive memory of what happened remains dissociated. In multiple personality disorder, which has been shown in 99% of cases to have developed as a self-protective response to repeated sexual and physical abuse in early childhood, two or more fragments of identity, sometimes called "ego-states" or "alters" are created, that maintain a continuity of their own, often with different names and different personality characteristics. As Hilgard says, "the concealed (or dissociated) personality is sometimes more normal or mentally healthy than the openly displayed one. This accords better with the idea of a split in the normal consciousness rather than the with the idea of a primitive unconscious regulated largely by primary process thinking."[29]

The notion of "splitting" of two or more equally rational and organized psychic fragments or identities was also used by Robert J. Lifton in his analysis of the Nazi doctors, who were able to enjoy listening to Beethoven in the garden and play with their children after a day of torturing and killing people. I believe that this concept of dissociation or splitting provides a more accurate and more useful understanding of the collective human pathology vis-á-vis the environment than the notion of a repressed and primitive "ecological unconscious." The entire culture of Western industrial society is dissociated from its ecological substratum. It's not that our knowledge and understanding of the Earth's complex and delicate web of interdependence is vaguely and inchoately lodged in some forgotten basement of our psyche. We have the knowledge of our impact on the environment, we can perceive the pollution and degradation of the land, the waters, the air—but we do not attend to it, we do not connect that knowledge with other aspects of our total experience. Perhaps it would be more accurate, and fair, to say that individuals feel unable to respond to the natural world appropriately, because the political, economic and educational institutions in which we are involved, all have this dissociation built into them. This dissociative alienation has been a feature of Western culture for centuries, or in some respects even for millennia, if Paul Shepard is right.

Elsewhere, I have argued that due to a complex variety of social and historical reasons, a core feature of the Euro-American psyche is a dissociative split between spirit and nature.[30] We have a deeply ingrained belief that our spiritual life, our spiritual practices, must tend in the opposite direction than our nature. Spirit, we imagine, rises upward, into transcendent realms, whereas nature, which includes bodily sensations and feelings, sinks or draws us downward. In some versions of this core image, the contrast between the two realms or tendencies is even sharper: there is not only a separation of the two tendencies, but incompatibility and opposition between them. The human spiritual is then always regarded as superior to the animal natural—and we are back with the humanist superiority complex described above.

Paul Shepard calls this the "central dogma of the West" and traces it to the Christian desert fathers. Susan Power Bratton on the other hand has argued that the desert fathers were the first ecologically conscious Christians.[31] Earlier fore-shadowings of this idea can be found in the Hebrew Old Testament, for example the following passage from Isaiah (55:8): "My ways are not your ways, says the Lord; just as heaven and earth are apart, so are my thoughts separate from yours."

Whenever this dissociative split originated, clearly by the time of the Protestant reformation, the idea is firmly implanted in almost everybody's mind that we have to overcome our "lower" animal instincts and passions, to conquer the body, in order to be spiritual and attain to "heaven" or "enlightenment." This image says: to enter into the city of God, the divine realms, you have to work against your nature (this was called the *opus contra naturam*). In the modern psychological, Freudian version of the ancient split, the conflict is between the human ego consciousness, which has to struggle against the unconscious body-based, animal id, in order to attain consciousness and truly human culture. Our conflicted relationship with the natural, what Freud called *Das Unbehagen in der Kultur*, the discontent of culture, was for him the price we had to pay for the possibility of civilization.

The similarity of the two formulations, the religious and the psychological, lies in this dualism: we could say that throughout the history of Western consciousness there has been a conception of two selves—a natural self, which is earthy and sensual, and tends downward, and a spiritual or mental self, which is airy and ethereal, and tends upward. Perhaps its most vivid formulation is by the 18th century German poet-philosopher Goethe, who formulated this core dualistic image in a famous passage in his drama Faust. "Two souls, alas, are dwelling in my breast, and one is striving to be separate from the other." One "holds to the world, with sensual, passionate desire"; the other "rises from earthly mist to the ethereal realms." The story of Faust, with his restless and ruthless quest for knowledge as personal power, strikes us as somehow a mythic key to the European psyche.

The ecologically disastrous consequences of this dissociative split in Western humans' identity become clear when we reflect upon the fact that if we feel ourselves mentally and spiritually separate from our own nature (body, instincts, sensations, and so on) then this separation will also be projected outward, so that we think of ourselves as separate from the great realm of nature, the Earth, all around us. If we believe that in order to advance spiritually we have to go against, to inhibit and control, the natural feelings and impulses of our own body, then this same kind of antagonism and control agenda will also be projected outward, supporting the well-known Western "conquest of nature" ideology. For most people in the West, their highest values, their noblest ideals, their image of themselves as spiritual beings striving to be good and come closer to God have been deeply associated with a sense of having to overcome and separate from nature.

It does not take much imagination to see how the consequences of this distorted perception have been played out in the spread of European civilization around the globe. And it is a distorted, counter-factual image: we human beings are not, in fact, separate or superior to nature, nor do we have the right to dominate and exploit nature beyond what is necessary for our immediate needs. We are part of nature—we are in the Earth, not on it. We are like the cells in the body of the vast living organism that is planet Earth. An organism cannot continue to function healthily, if one group of cells decides to dominate and cannibalize the other energy systems of the body.

Furthermore, the idea that the spiritual and the natural are opposed or that spirituality must always transcend nature is a culturally relative concept not shared by non-monotheistic religions or traditional societies. In indigenous cultures around the world the natural world is regarded as the realm of spirit and the sacred; the natural is the spiritual. From this follows an attitude of respect, a desire to maintain a balanced relationship, and an instinctive understanding of the need for considering future generations and the future health of the ecosystem—in short, sustainability. Recognizing and respecting worldviews and spiritual practices different from our own is perhaps the best antidote to the West's fixation in the life-destroying dissociation between spirit and nature.

Notes

1. Wilhelm Reich, *The Mass Psychology of Fascism* (New York: Orgone Institute Press, 1946).
2. Warren M. Hern, "Why Are There so Many of Us? Description and Diagnosis of a Planetary Ecopathological Process," *Population and Environment*, vol. 12, no. 1, 1990.
3. Theodore Roszak, *The Voice of the Earth* (New York: Simon & Schuster, 1992).
4. Al Gore, *Earth in the Balance: Ecology and the Human Spirit* (Boston: Houghton Mifflin Co., 1992), 217.
5. James Lovelock, *Healing Gaia: Practical Medicine for the Planet* (New York: Harmony Books, 1991), 154.
6. George Sessions, "Ecocentrism and the Anthropocentric Detour," *ReVision*, vol. 13, no. 1, Winter 1991.
7. Gary Snyder, *The Practice of the Wild* (San Francisco: North Point Press, 1990).
8. David Ehrenfeld, *The Arrogance of Humanism* (New York: Oxford University Press, 1978).
9. Ralph Metzner, "The Split between Spirit and Nature in European Consciousness," *The Trumpeter*, vol. 10, no. 1, Winter, 1993.
10. Paul Shepard, "A Post–Historic Primitivism," in Max Oelschlaeger, ed., *The Wilderness Condition: Essays on Environment and Civilization* (San Francisco: Sierra Club Books, 1992), 85.
11. Ibid., 85.
12. Paul Shepard, *Nature and Madness* (San Francisco: Sierra Club Books, 1982), 16.
13. Ibid., 40.
14. Erik Erikson, *Identity and the Life Cycle* (New York: Norton & Co., 1959; 1980).
15. Ibid., 97.
16. Ibid., 98.
17. Shepard, "Primitivism," 85.
18. Jean Liedloff, *The Continuum Concept* (Reading, Mass.: Addison–Wesley, 1975).
19. Shepard, *Nature and Madness*, 124.
20. Ibid., 124.
21. Ibid., 128.

22. Thomas Berry, "The Ecozoic Era," *Eleventh Annual E.F. Schumacher Lectures, 1991* (Great Barrington, Mass.: E.F. Schumacher Society, 1991).
23. American Psychiatric Association *Diagnostic and Statistical Manual of Mental Disorders* (DSM–III–R), (Washington, DC: APA, 1987).
24. Dolores LaChapelle, *Sacred Land, Sacred Sex* (Durango, Col.: Finn Hill, 1988), 48.
25. Paul Devereux, *Earthmind* (New York: Harper & Row, 1989), 2–3.
26. Immanuel Velikovsky, *Mankind in Amnesia* (Garden City, N.Y.: Doubleday & Co., 1982).
27. Roszak, *Voice*, 320.
28. Ernest Hilgard, *Divided Consciousness* (New York: John Wiley & Sons, 1986).
29. Ibid., 83.
30. Metzner, "The Split."
31. Susan Power Bratton, "Sleeping with Lions: The Wild and the Holy," *Orion*, Summer 1991.

Play and Place in Human Development

Dolores LaChapelle

> You cannot go back. You cannot turn the clock back. You cannot become primitive people—that's all behind us. We have to face the modern world in its own terms—angst and all that Western Philosophy has given us. But, that's only true if our culture is always created—creates itself. It cannot be true if an organism is involved.... If we are talking about human needs, rather than needs generated by a particular ideological view, then you can't go back, because you haven't left. You aren't going back to something primitive. You are recovering an essential process in maturation which should take place within the person at sixteen years or whatever.[1]

I MET PAUL SHEPARD IN 1977 in Claremont, California, at the planning meeting to begin the New Natural Philosophy Program for International College in Los Angeles.

Shepard lives in a very small house up one of the canyons out of Claremont. There he keeps jungle fowl in order to study their rituals. One night Paul had a party so that I could meet his friends and colleagues. The first group arrived in time to see the evening ritual. Jungle fowl, the ancestors of modern chickens, can still fly into trees. Paul keeps the nesting females inside a wire enclosure as protection against the small predators of the area. The sun was off the canyon and it was getting toward twilight. The cock who lives up in the trees, flew down, went to each of the hens in his harem and gave a long speech. That ritual closing of the day accomplished, he flew back into his tree and want to sleep. Sometime later another group of people arrived. The car headlights woke the cock up; he again flew down, went from hen to hen in the same ritualized manner, and then returned to his roost for the night.

This event perhaps shows why Paul Shepard's work is unique. He is not only an outstanding scholar and teacher, but for most of his life he has been in close contact with the wild—the land and the creatures—wherever he lived: thus he knows information directly from the animals, knowledge that can not be secured in any other way. Such learning takes time and dedication. Shepard's work draws together the research of ethologists, such as Konrad Lorenz and Tinbergen, together with the findings of the best of the anthropologists. Furthermore, making use of the conclusions from the Man the Hunter Conference in 1966, he has clarified the crucial role of the hunter and forager in human development. "Cultural Man has been on earth for some 2,000,000 years; for over 99 per cent of this period he has lived as a hunter-gatherer. Only in the last 10,000 years has man begun to domesticate plants and animals."[2] Thus we have been the same human beings for at least 50,000 years.

A succinct example of the unique breadth and clarity which Shepard brings to his work on human development can be found in a brief passage in *Man in the Landscape*, where he explains the human eye. "The sea eye was first connected to a fishy brain and body. When taken on land it continued to see only the spectrum of light transmitted by water and to commutate the 'coolness' of the blue depths and the 'warmth' of the reddish, sunlit surface in a hundred ways." And later, he writes:

> The bush apes ancestral to man gradually worked farther from the forest. They took the sea eye out of the gloaming and into the radiance of the open day, though not without some enduring nostalgia. The awareness of glare in the open air and a sense of vulnerability when in it have not been completely dispelled. An affinity for shade, trees, the nebulous glimmering of the forest interior, the tracery of branches against homogenous surfaces, climbing, the dizzy childlike joy of looking down from a height, looking through the windows and into holes, hiding, the mystery of the obscure, the bright reward of discovered fruits are all part of the woody past.[3]

Just as the embryo recapitulates the stages of evolution within the womb, the developing child must go through the necessary stages to become an adult. There are time-critical constraints and needs. Shepard argues that

> The whole of growth through the first twenty years is our "coming into being," or ontogeny.... Among those relict tribal peoples who seem to live at peace with their world, who feel themselves to be guests rather than masters, the ontogeny of the individual has some characteristic features. I conjecture that their ontogeny is more normal than ours . . . and that it may be considered to be a standard from which we have deviated. Theirs is the way of life to which our ontogeny was fitted by natural selection, fostering a calendar of mental growth, cooperative leadership, and the study of a mysterious and beautiful world where the clues to the meaning to life were embodied in natural things, where everyday life was inextricable from spiritual significance and encounter, and where the members of the group celebrated individual stages and passages as ritual participation in the first creation.[4]

In the childhood phase of ontogeny the most important activity is play. Shepard points out: "The 'founding' of a world in childhood and the special mystical experience related play and ritual.... The special places of childhood are not sacred, but the memory of them is necessary for attaching sacredness to place."[5] Essentially ritual grows out of play; thus ritual has been called "deep play."

Shepard reprinted Edith Cobb's only essay on her life-long research into the importance of play in middle childhood in his edited book, *The Subversive Science: Essays Toward an Ecology of Man* in 1969. Her work has been a continuing influence on him; when I first met Paul at Claremont, he handed me her book (just published) and told me of her importance. Cobb spent her life researching and studying primarily one thing: What is it about childhood that makes that child into a genius? Along with her other research she gathered together biographies and autobiographies of 300 creative people from Africa, Asia, Europe and the Americas dating back as far as the Middle Ages. What she found in common among them was an experience in middle childhood she called a "nature mystic experience, where the natural world was experienced in some highly evocative way, producing in the child a sense of some profound continuity with natural processes."[6]

Consider two descriptions of this "nature mystic experience." The noted poet Kenneth Rexroth describes it as "An awareness of timeless, spaceless, total bliss. . . . This seemed to be my normal and natural life which was going on all the time and my sudden acute consciousness of it only a matter of attentions."[7] Bernard Berenson writes that "In childhood and boyhood this ecstasy overtook me when I was happy out of doors. Was I five or six? It was a morning in early summer. A silver haze shimmered and trembled over the lime trees. The air was laden with their fragrance. I remember...that I climbed up a tree stump and felt suddenly immersed in Itness. I did not call it by that name. I had no need for words. It and I were one.[8]

Talking about Cobb's studies on genius, during a lecture at the University of Utah, Shepard explained: "One of the things that she found was that at various times in their life, particularly when they had a problem, they [the geniuses in her study] went back to where they were as children and walked through the landscape.... They walk through in some strange tutelary way: this re-evoked and reanimated and revivified their own genius of the place."[9]

Recent research into shamanism has uncovered an even deeper level of the importance of place in childhood. The German anthropologist Ruth Inge-Heinz, who has studied shamanism in Southeast Asia, Nepal and Tibet, hosts a yearly conference on the subject. In the introduction to her book, *Shamans of the Twentieth Century*, she writes: "This book has been on my mind for a long time. Over sixty-five years ago, when I was of pre-school age, I connected with everything around me—the sun, the wind, the sea, the earth. This experience at the shores of the Baltic Sea was truly ineffable. I kept it to myself and, in times of despair, I would recall the infinite love and energy I had felt pulsing through me at that time." Further, she says "I survived the dark years (1933-1945) in Germany, which included two-and-a-half years of starvation and bombing in Berlin, by drawing from inner resources.... Remembering my childhood experience during which I had met the infinite source, I learned to access this source every time I needed nourishment."[10]

In a similar vein, C. Jess Groesbecks, a Jungian analyst, says that when the shaman journeys to the "other world," he is actually visiting his childhood landscape.[11]

In traditional cultures, the crucial role of the place itself began when the child was born. As an example, consider this chant from the Omaha tribe:

> Ho! Ye Sun, Moons, Stars, all ye that move in the heavens. . . .
> Ho! Ye Winds, Clouds, Rain, Mist, all that move in the air. . . .
> Ho! Ye Hills, Valleys, Rivers, Lakes, Trees, Grasses, all ye of the earth. . . .
> I bid you hear me!
> Into your midst has come a new life! Consent ye, I implore.[12]

Within a few days after the child's birth, by means of this ritual chant, the Omaha medicine man asked permission of all the other beings of the tribal place for the child to come into that place. This ritual acknowledgment externalized the knowledge, which tribal peoples had, that the human being had no special claim on the right to birth but, instead, must make every effort not to upset the balance of the total environment—that is, the place. Living wholly within a specific locality so-called primitive humans had learned, as Gregory Bateson observes, that "the organism which destroys its environment destroys itself."[13] For primitive hunter-gatherers the birth of a child was not an accident, left up to the whim of individual parents; rather birth was regulated by ritual, so that the particular child would fit into the overall health and stability of the entire place—the tribe plus the rest of the community of soil, animals and plants. In such cultures the emphasis was on encouraging the child to be aware of all nature. Total awareness of place was a necessity; the hunter must know the hidden spots where game abounds and the route the game uses to reach water. Gatherers must know where the choice roots grow and identify food plants by the shape of leaves. (Some tribes recognized as many as 1200 species of plants by sight.)

More generally, for ninety-nine percent of the time humans have been on this earth they were hunter-gatherers, people who maintained responsible relationships to particular localities. Experts agree that there must have been systematic birth control practices as long ago as the Pleistocene to account for the difference between potential and real growth rates during that era. Today the earth is being overwhelmed by a growing mass of humans. Shepard argues, in his book *The Tender Carnivore and the Sacred Game*, that human overpopulation is an artifact of the collapse of the Pleistocene economy. "The collapse of an ecology that kept man scarce and attuned to the mystery and diversity of all life led, as though by some devilish Fall, to the hunting and herding of man by man, to the hoarding of grain and the secularizing of all space. To defend his fields a farmer needed many kinsmen, sons, and co-defender, and co-fighters, and ultimately brother ideologists."[14]

CHILDHOOD PLAY AND ADULT RITUAL

The importance of place in the relationship between childhood play and adult ritual can be illustrated by two ceremonies recorded by early ethnologists.

(1) "Turning the Child" Ceremony (Omaha)

When the child could walk steadily by itself, about the age of three or four, the Omaha would hold this ceremony in the Spring, after the first thunder was heard and the meadow larks were singing. Peals of thunder were regarded as signals indicating the awakening the life-giving forces from the sleep of Winter. In the course of the ritual the child was symbolically "sent into the midst of the winds." A special tepee was set up and "made holy." A stone was placed in the center near the fire. Sweet grass was laid upon it. The child's parents would bring, along with the child, a pair of new moccasins to the ceremony.

The mother would lead the child up to the open door and ritually speak with the shaman within, saying: "Venerable man! I desire my child to wear moccasins." Then she dropped the child's hand; the little one, carrying the new moccasins, entered the tepee alone. The mother then said, "I desire my child to walk long upon the earth; I desire him to be content with the light of many days." The priest would reply, addressing the child: "You shall reach the fourth hill [old age] sighing; you shall be bowed over; you shall have wrinkles.... I speak to you that you may be strong." Then, laying his hand on the child's shoulder, he would lead the child to the fireplace in the center of the lodge, saying in his capacity of Thunder, whose priest he was: "I am a powerful being; I breathe from my lips over you." Then the shaman would chant the Invocation to the Winds:

Ye four, come hither and stand, near shall ye stand.
In four groups shall ye stand.
Here shall ye stand, in this place, stand.
(The Thunder rolls)

At the close of the ritual chant, the priest turns the child toward the east and lifts it by the shoulder onto the stone. He then turns the child completely around from left to right and then set the child back down on the rock, now facing south. The priest lifts the child again, turning it all the way round, setting it down facing west. Then, for a third time he lifts the child, turning it and setting it down facing north. The last time the child is turned again until it faces west. During this ceremonial process, the following chant is sung:

Turned by the wind goes the one I send yonder;
Yonder he goes who is whirled by the wind;
Goes where the four hills of life and the four winds are standing;
There, in the midst of the winds do I send him.
Into the midst or the winds, standing there.
(The Thunder rolls)

(Alice Fletcher and the Omaha, Francis La Flesche, who together wrote the Bureau report on this ritual, relied on the direct testimony of an old man. The old man, as a curious child, had hidden himself in the Sacred Tent to witness the ceremnony. Fletcher and LaFlesche explain that "The winds invoked by the priest stand in four groups, and receive the child, which is whirled by them, and by them enabled to face in every direction." The informant said that this "brought the child face to face with the life-giving Winds of the four directions.")

The priest next put the new moccasins on the feet of the child, and then sets the child on its feet to take the four ritual steps—the beginning of its entrance into a long life. The new name of the child was announced and the priest cried aloud: "Ye hills, ye grasses, ye trees, ye creeping things both great and small, I bid you hear! This child has thrown away its baby name. Ho!" Thus the small child is sent out from its childish dependence on the complete authority of its parents into the greater world of Grandfather Rock and Grandmother Earth and the four sacred winds and the Thundergod. The child learns to respect and love these nonhuman beings with the result that it is not subject to the limitations of human relationships alone.[15]

To begin to understand the depths of this ritual, it is necessary to grasp some of the meanings of the rock and the winds and thunder in the Omaha culture. Their tribal world was bounded by the circle of the horizon, "the rim of the world about which walk the winds." In the Plains Indians' life, the winds served as a unifying principle because they move and exert power over the forms of nature—the unifying element common to the sacred elk, the female elk, the birds, and the sacred cottonwood tree—but are in themselves unseen. The whistling call of the elk, caused by the use of his breath, attracts the cow to him and thus brings about the birth of young. Birds are called those-who-walk-upon-the-wind. The cottonwood tree sends out its seeds enclosed in downy cotton filaments, so the seeds can ride the wind. Part of the sacredness of this tree is due to the fact that its presence indicates water near the surface of the ground. The sacred pole used for the Sun Dance is always a cottonwood tree.

The "Turning the Child" ritual was only the beginning of the relationship between child and Nature. Not only was the child symbolically sent out into the midst of the winds, but at important times in its young life the child was encouraged to actually go out into Nature and let "the Powers" decide the outcome. The psychologist, Abraham Maslow, writes of an incident (which occurred when he lived on the Blackfoot reservation in the 1920s) illustrating the point. A small Indian child of seven owned a medicine bundle of great value. When an Anglo museum official offered to buy it from him, the child went off into the wilderness by himself to meditate. He was gone two or three days and nights. Maslow says that he did not ask his father or mother for advice; neither did they offer any. He came back and announced his decision, which was not to sell the medicine bundle; they never questioned that either, since the child belonged just as much to the great powers of Nature as he did to them. Maslow says, "I can just see us doing that with a seven-year-old-kid."[16]

Rituals, such as "Turning the Child," involving interactions between the child and natural objects, make for a more comprehensive learning than is possible by means of our culture's more limited linear-analytical-left-hemisphere of the brain approach to learning. According to the ecologist, Eugene Odum, the living matter all around, the biomass, is stored information. He says that there is more information of a higher order of complexity stored in a few square yards of forest duff than there is in all the libraries of the world. It is a different order of information, because it is the flow of information of billions of years of life on this earth, before the appearance of Human consciousness (itself limited to the human purposes).[17] The natural world has forms of consciousness with different purposes than merely human ones; arguably, interaction with these other forms of consciousness expands the human mind.

Actually, the human brain consists of three brains—each with its own special intelligence, memory, and other functions: the reptilian brain, which evolved several hundred million years ago, the limbic or mammalian brain, and the outermost or newest brain, the neo-cortex. The reptilian brain plays a role in aggressive behavior, territoriality, and ritual. The mammalian brain has to do with emotions, relationships, and smells among other functions. The mind as a whole is an integrated network of circuits going throughout all three levels, but the conscious mind registers only the bits and pieces of the other circuits. But consciousness is not limited to the neo-cortex alone. The sudden flood of sensations and memories triggered by certain smells is an indication of the interaction of the older brains with the new. Most of the activities of our culture cut us off from the older brains. Ritual actions with natural beings tend to restore communication between the three brains and the non-human forms of consciousness, that is, the plants and animals.

(2) "Dance of the Hasjelti" Ceremony (Navajo)

A second example of education which bonds the child to its place comes from the Navajo. Their "Dance of the Hasjelti" provides one of the clearest examples of education which is not limited to the linear function of the left-hemisphere, but instead includes complete body/brain interaction. This ritual was witnessed and recorded in 1885, when ceremonial life was still relatively intact. The Dance of the Hasjelti, held on the eighth day of the nine-long Night Chant, was for initiating boys and girls into the tribe's ceremonial life.

According to Navajo mythology, Hasjelti and Hostojoboard were the children of Aksonnutli, "the turquoise" and "white shell woman." The turquoise placed an ear of white corn and white shell woman placed an ear of yellow corn on "the mountain where fogs meets." The white corn conceived and the child was Hasjelti; the yellow corn conceived and the child was Hostojoboard. Born on the Henry Mountains in Utah, they travelled to each of the four sacred mountains of the Navajo: Sierra Blanca in Colorado, San Matzo in New Mexico, San Francisco in Arizona, and Ute Mountain in Colorado, giving two songs and prayers on each mountain. Thus these two sacred beings were called the great song makers of the Navajo world. Prayers for rain and snow are made to them. Hasjelti mediates between the people and the sun and communicated with the people by means of the feathers and plumes of birds, "Those who walk on light."

At the beginning of the ceremony the children were brought to the mesa where the medicine lodge was located. A buffalo robe, with the head to the east, was spread at the end of the space which extended out from the lodge. Blankets were spread in a semi-circle around the robe. After being stripped of their clothing, the boys sat with bent heads on the buffalo roles. The girls, with their heads bent down, but dressed in blankets and jeans, sat in line upon the blankets.

At this particular ceremony in 1885, nine boys and six girls were initiated. They were all between five and ten years of age; up to the actual time of the ceremony they had never seen any of the ritual implements nor the masks. When all were in position, Hasjelti, wearing black velvet ornamented with silver and carrying a fawn skin containing sacred meal, and Hostjoboard, her nude body painted white, with a silver belt holding a scarf around her loins and carrying two needles of the Spanish bayonet plant, stood in front of the children. The first boy was led out and stood facing east the direction of the sun's rising. Hasjelti drew a cross on the boy's back with the sacred meal and gave his peculiar hooting sound. Hostjoboard then struck the boy upon the chest, first with the needles held in her right hand, and then with those held in her left hand. Hasjelti turned the boy toward the right until he faced west, where the sun sets, and made a cross on his back. Hostjoboard struck him twice with the needles. He was again turned to face east. Both arms were extended and brought together. Hasjelti made a cross over the arms and then over the knees. Each time Hostjoboard struck the spot with the needles. Then the boy returned to his seat. When all the boys had taken part, the sacred beings moved down the line of the girls. Hasjelti marked a line of meal on each side of the foot of the girl. Hostjoboard, now holding two ears of yellow corn wrapped with pi§on twigs, placed them to the soles of the girls feet. Hasjelti drew a line of meal on each hand, followed by Hostjoboard, who placed an ear of corn to the palms of each hand. Hasjelti formed a cross on the chest with the meal, and Hostjoboard pressed the corn to the chest. A cross was also made on the back, followed by pressing corn to the back. With his right hand, Hasjelti drew a line on the girl's left shoulder and, with his left hand, a line on the girl's right shoulder. The corn was also pressed to each shoulder in a similar manner. Two lines of meal were run over the forehead, back to the top of the head, and the ears of corn were pressed to the top of the head.

At the close of the ceremony, the representatives of the gods removed their masks, and told the children to raise their heads. "The amazement depicted upon the faces of the children when they discovered their own people, and not gods, afforded much amusement to the spectators," according to the report. The masks were laid upon a blanket, and the children were told to look at them. Then Hostjoboard placed her mask upon the face of each boy and girl in the line, giving a hoot each time and taking care that the child's eyes looked out directly through the eyeholes. Then the mask was laid beside that of Hasjelti's and both masks were sprinkled with pollen by the representatives of the gods. The boy at the end of the line was then called out. He took a pinch of pollen from the bag and sprinkled each mask. This was repeated by each person who was present in a ceremonial fashion: a line of pollen was run from the top of the head down to the mouth. The younger children's hands were carefully guided by the representatives of the gods. The line running down through the center of the face called upon the gods to send down rain upon the earth and health to all the people. The lines running up the cheeks were for the vegetation that grows upward from the earth.

On the following day there was a great dance for all the people. In this dance the new, initiated children took their part in the tribal ceremony, now knowing that they, too, were part or the great interrelationship among gods and nature and the human beings of the tribe in that place.[18]

To begin to understand the educational aspects of this ritual, it is important to remember that the two older brains are non-verbal; communication involving these brains must take place by primary process rather than merely verbal methods. Primary process works through such methods as movement, dance, art, dreams and symbols. Long ago, when Isadora Duncan (the great dancer) said, "If I could tell you what it meant, there would be no point in dancing it," she unwittingly explained the paradox of primary process communication.

For the Navajo, pollen is a symbol for life and productivity. It symbolizes light as well. It is easy to trace the origins of these ideas back to their cultivation of corn. Corn has to be cross-pollinated. The pollen from the male cells in the tassel has to be deposited on the sticky corn silk of another plant in order for an ear of corn to be produced. In one sense pollen is nearly immortal. Each tiny grain is protected by a very tough plastic-like covering, smooth as glass, which lasts practically forever when protected from air. This covering withstands high temperatures and all acids. When buried in mud, pollen becomes a microfossil. Archaeologists date ancient ruins by means of pollen. All of these factors go into the symbolical use of pollen.

For the Navajo, corn is synonymous with human life. Indian culture and corn seem to have developed—co-evolved—together. Corn is so ancient that experts today cannot find the native plant from which it evolved. This ancient relationship between humans and corn is implied in the descent of Hasjelti and Hostjoboard from white and yellow corn. Hasjelti's mask has a stalk of corn right down the middle of the face. Hostjoboard placed ears of corn to the hands and feet of each girl. The boys and girls are marked with the sacred meal, symbolizing the life brought to them from corn. The stalks of the Spanish bayonet (a stiff, rough-edged plant) used to whip the boys symbolized the dangers and pain of the hunting life. Together, these two symbols, meal and the Spanish bayonet, symbolize life and death, health and suffering, which nature brings to humans.

The children were marked with a cross of meal, thus indicating that the lines of the sacred four directions of that place met here in each child. When the gods remove their masks, the children see that their own relatives are conduits for the power of the gods, which is coming into the children. Then the god's masks are put on the children, clearly indicating that they, too, will grow up to assume some of the power of the gods and, in turn, pass it on. The children sprinkle pollen, the life-giving symbol, on the masks, indicating their role in continuing tribal life. Before, in their daily life, they playfully took the part of animals or trees or corn plants. In this ritual, their play world joined with the adult world of tribal ceremony as each child experienced the fact that he or she was a part of the ongoing drama of interaction between plants, humans, gods and earth in that particular place.

CONCLUSION

Not only in hunting-gathering cultures, but also in the ancient classical cultures, such as that of the Greeks, play was considered sacred. The Olympics began as games for the gods. This link between play and worship has been succinctly explained by Paul Shepard, who writes that "ritual is always a signal that condenses (or symbolizes) information about a relationship. . . . The 'as if' of play is the heart of ritual and, eventually, formal religious activity." Later, in formal rituals, "one moves between bases as in a game, now remembering that it is really Jim Smith behind the mask in the ceremony and now suspending that knowledge in transcendental experience. Play and worship are the two human activities removed from commonplace life. One is necessary for the other."[19]

In a study of such adult play activities as mountain climbing and marathon swimming, Mihaly Csikszentmihalyi developed a definition of play as "the state in which action follows upon action according to an internal logic which seems to need no conscious intervention on our part. We experience it as a unified flowing from one moment to the next, in which there is little distinction between self and environment; between stimulus and response; or between past, present and future."[20] This experience happens only when a person is actively engaged in some form of playful interaction with the environment that leads to a merger of action and awareness. Or in ritual, where ceremonies create, by design, the merger of body and mind with the natural surroundings. As soon as there is a split, so that the conscious mind stands outside the action, looking on, the flow and thus the possibility of a merger is lost. The rational, linear left hemisphere takes over, interrupting the flow of communication between the rest of the brain and the body. It follows that the play of childhood is a necessary biological prerequisite for adult cultural rituals.

In traditional cultures that practice rituals bonding child and nature, there is no adolescent crisis. The adolescents in some cultures go on a vision quest, but this is not a personal crisis. It is a seeking from the greater beings of the place for guidance in how that person can best serve the tribe and the place. In contrast, Shepard explains what happens to the adolescent male in our culture: "The fiction granted him by the pseudopastoral desert philosophy of the West is that his painful incompleteness is the true mature experience and that the meaninglessness of the natural world is its meaning. . . . Acted upon, it wounds us and we wound the planet." Thus the young man's normal human development is arrested; "denied the mythopoetic visions of man in nature, he will for the rest of his life struggle with existential problems that are normally the work of a few critical years in his second decade of life." Shepard argues that a basic "framework of nature" is needed in the adolescent years as much as a nutritious diet. "Lacking it, he will always lack true reverence for the earth. The remaining choices for a logic of creation are an otherworldly orientation, materialist exploitation, or existentialist absurdity."[21]

Near the end of his book, *Nature and Madness*, Shepard sums up by asking: "What can one say of the prospect of the future in a world where increasing injury to the planet is a symptom of human psychopathology?" The situation is more hopeful than we might think if we understand that

> an ecologically harmonious sense of self and world is not the outcome of rational choice. It is the inherent possession of everyone; it is latent in the organism, in the interaction of the genome and early experience. The phases of such early experiences, or epigenesis, are the legacy of an evolutionary past in which human and nonhuman achieved a healthy rapport. Recent societies have contorted that sequence.

He continues by saying: "Perhaps we do not need new religious, economic, technological, ideological, esthetic, or philosophical revolutions. . . . The civilized ways inconsistent with human maturity will themselves wither in a world where children move normally through their ontogeny."[22]

The final paragraph of *Nature and Madness* offers a fitting conclusion to my essay:

> The problem may be more difficult to understand than to solve. Beneath the veneer of civilization, to paraphrase the trite phrase of humanism, lies not the barbarian and animal, but the original human who knows the rightness of birth in gentle surroundings, the necessity of a rich nonhuman environment, play at being animals, the discipline of natural history, juvenile tasks with simple tools, the expressive arts of receiving food as a spiritual gift rather than as a product, the cultivation of metaphorical significance of natural phenomena of all kinds, clan membership and small-group life, and the profound claims and liberation of ritual initiation and subsequent stages of adult mentorship. There is a secret person undamaged in every individual, aware of the validity of these, sensitive to their right moments in our lives. All of them are assimilated in perverted forms in modern society: our profound love of animals twisted into pets, zoos, decorations, and entertainment; our search for poetic wholeness subverted by the model of the machine instead of the body; the moment of pubertal idealism shunted into nationalism or ethereal otherworldly religion instead of ecosophical cosmology.[23]

Notes

1. Paul Shepard. Lecture at University of Utah, Salt Lake City, Utah. February 13, 1985.
2. Richard B. Lee and Irven DeVore, eds., *Man the Hunter* (Chicago: Aldine Publishing Co., 1966), 3.
3. Paul Shepard, *Man in the Landscape: A Historic View of the Esthetics of Nature* (College Station, Tex.: Texas A & M University Press, 1967; 1991), 22.
4. Paul Shepard. *Nature and Madness* (San Francisco: Sierra Cub Books, 1962), 6.
5. Shepard, *Man in the Landscape*, 37.
6. Edith Cobb. "The Ecology of Imagination in Childhood," *Daedalus* 88 (Summer 1959):537-548.
7. Kenneth Rexroth, *An Autobiographical Novel* (New York: New Directions, 1969).
8. Quoted in Edith Cobb, *The Ecology of Imagination in Childhood* (Dallas: Spring Publications Inc., 1993), 32.
9. Shepard, Lecture 1985.
10. Ruth Inge-Heinz, *Shamans of the Twentieth Century* (New York: Irvington Publications Inc., 1991), 18-20.
11. C. Jess Groesbeck, Lecture for Julien Puzey's class on Deep Ecology, Continuing Education program in Salt Lake City, Utah, 1965.
12. Alice C. Fletcher and Francis La Flesche, "The Omaha Tribe," *Twenty-seventh Annual Report* (Washington, D.C.: Bureau of American Ethnology, 1911).
13. Gregory Bateson, "Pathologies of Epistemology," *Second Conference on Mental Health in Asia and the Pacific*, 1969 (Hawaii: East-West Center Press, 1972). Reprinted in *Steps to an Ecology of Mind.*
14. Paul Shepard, *The Tender Carnivore and the Sacred Game* (New York Charles Scribner's Sons, 1973), 7.
15. Fletcher and La Flesche, "The Omaha Tribe."
16. Abraham Maslow, *The Farther Reaches of Human Nature* (New York: Viking Press, 1971), 231.
17. H. T. Odum, *Environment, Power and Society* (New York: John Wiley, 1971), 172.
18. S. Stevenson, "Navajo Myths," *Eighth Annual Report* (Washington, D.C.: Bureau of American Ethnology, 1889). Depending on the informant and the time the material was gathered, the locations and names of the sacred mountains vary.

19. Shepard, *Tender Carnivore*, 196, 199.
20. Mihaly Csikszentmihalyi, "Play and Intrinsic Rewards," *Journal of Humanistic Psychology* (Summer 1985):44- 63.
21. Shepard, *Nature and Madness*, 70-71.
22. Ibid., 128-129.
23. Ibid., 129.

Bios, Mythos, and Arid Zones: An Appreciation of Paul Shepard's Critique of the Desert Fathers

Carl H. Hertel

> In the domain of myth is to be found not ordinary truth but a higher truth, which permits approaches to itself from the domain of bios.
>
> —C. Kerenyi (quoted in *The Sacred Paw*, 135)

ANYONE WHO HAS READ Paul Shepard's *Nature and Madness* knows how he feels about the 4th Century Christian Desert Fathers and their Hebrew predecessors as regards themselves and the arid part of the biosphere they inhabited. According to Shepard, they did not value the desert. "It was a vacuum, idealized as a state of disengagement and alienation, a symbol of the condition of the human spirit." As luck would have it, it fell to me during twelve years of team-teaching a course called "The Desert as a Place" at Pitzer College during the 70s and 80s with Paul Shepard to introduce the Desert Fathers and defend them and their spiritual pursuits as best I could against his characterizations. It was somewhat like setting up ducks in a shooting gallery as I sought to make a case for an "ecological spirituality" around the "wisdom and emptiness" model I had adopted. In retrospect, I see that I should have sought more compelling examples, ones with demonstrable roots in bios as well as a home in mythos. Thereby, I might have attempted to set up direct bio-physiological links between these desert inhabitants and their cultural and spiritual behaviors. In short, reading the Verba, and intimating vague ties to the Chan/Zen of Bodhidharma resulting from his trek across the Taklamakan and Gobi, could not cut it when it came to Shepard's analysis of Near Eastern deserts and the roots of Western culture's ecological myopia. In his analysis he explores the ravages of a historical view of the world, biophobia and human alienation transposed from the deserts to northern European forests and ultimately to the childish "American Bedouin" of Lapham characterized by, in Shepard's words, "a sort of amalgam of teenybopper frothiness and macho juvenile vapidity clearly not synonymous with the seriousness of the Hebrew desert fathers." (But, I might add, even more extreme in their lack of concern for the land and in their violent trashing of habitats.)

These days I would seek a better balance of bios-mythos in my examples for these discussions. For instance, I might resort to a discussion of the work of contemporary scientist and environmental sculptor, James Turrell, whom I have discussed in a piece titled "Musing on Roden Crater: Near West Mountain, Home of White Corn Boy and Yellow Corn Girl, Fastened to Earth With a Sunbeam." The substance here is light. Turrell works with atmospheric effects such as available light (night and day), celestial vaulting, neural-optical occurrences such as Ganzfelds, and numerous other psycho-physiological responses of the human organism to light in "situations" in nature enhanced by his interventions (environmental sculpture such as the carefully leveled lip of his Roden Crater in Northern Arizona). It is in the bio-physical realities of light itself (in this case in a desert environment resulting from volcanism) that Turrell's work is to be both apprehended and understood. It is a case in which both bios and mythos play significant roles. Thus we may say the following about the desert and the psycho-spiritual dimensions of Turrell's work[1]: "Essence and life are invisible. So they are associated with sky and light. Sky and light are invisible, so they are associated with the two eyes" (*The Secret of The Golden Flower*). As Thomas Cleary says, in his translator's introduction to the above text, "The golden flower symbolizes the quintessence of the paths of Buddhism and Taoism. Gold stands for light, the light of the mind itself; the flower represents the blossoming, or opening up, of the light of the mind..."

This may be a long way from Ricky Skaggs' "The Highway 40 Blues," but right on the mark for Jim Turrell's work at Roden Crater outside of Flagstaff just north of Highway 40.

The "light" of which we speak lies at a point between the sky and the mind, but not necessarily on a line between the two—if indeed we can say something like "between" the sky and the mind, or for that matter between the sky and the brain, or between...sky and eye.... So, we have a unique locus or element of space which can also be said to be "invisible" like essence, life, sky and light (above): likewise space.

Ricky Skaggs' country Western wisdom begins to seem more relevant, keeping us down to earth—blues, highways, "lights of home," etc.

Ming, a Chinese character meaning "light," is composed of the ideograph for sun (*ri*) and moon (*yue*). This seems to require little explication. Juxtaposed as two individual characters (i.e., as *ri* and as *yue*) they mean "life"—all things that exist between night and day. Thus the plot thickens.

To quote Skaggs once again: "Lord I paid my dues" counting the days in a garden of volcanic flowers—"I got them Highway 40 blues. . . ."

If Turrell's pieces "work" we get somewhere beyond the sun and the moon. Thus, an essence—or as the Chinese say, *Qi* (pron. "chee") comes to play in the interface between the desert environment and the work. Other than "vital breath"—which we sometimes hear as a translation—*Qi* resonates with notions such as the Sanskrit *anahata nada*—the inaudible vibration—or *spandana, akarshana*—the essential vibrational-magnetic forms involving (perceptually speaking) only "space"—thus invisible-inaudible, but the basis for all perception. Hence, we have arrived at Shivaism and find ourselves in the Kali Yuga. Himalaya dreamscapes of (west) Tibet. Turrell seems to be looking down from the sixties when he flew missions mapping that terrain as a part of intelligence efforts.

It seems amazing to me you can do all this with only an old volcano, sky, sun, moon, stars, eyes, brains, skin, sweat, Turrell's Qi and, of course, money. As to site, we are not confined to high desert plateaus for this work. But I favor deserts; aesthetically, one values the fact that many religious traditions had their beginnings in desert areas—such ecosystems do something to the "eye," those long horizons of earth and sky—phosphene generators—ganzfield galvanizers—trigger the notion of extremes and the need for paying attention in order to survive as human animals—infinitesimal mineralities, the dryness—the limitable dryness, all that "we" are not, our perfect "other"—so, we do pay attention—all that clarity and aloneness.... Ribh (the hermit) calls it up: "these eyes /By water, herb and solitary prayer/ made aquiline, are open to that light."

Thus, sing it! Ricky, play that thing!...Outside Winona, head north away from "these Highway 40 blues.... I've walked holes in both of my shoes...counted the days since I been gone—and Lord, I'd love to see the lights of home..." Meanwhile, nearby...

Speaking of volcanic outcropping near Leupp, Arizona, Joy Harjo reflects in her *Secrets from the Center of the World*: "If all events are related, then what story does a volcano erupting in Hawaii, the birth of a woman's second son near Gallup, and the shoulder bone of earth made of a mythic monster's anger construct? Nearby a meteor crashes. Someone invents aerodynamics, makes wings. The answer is like rushing wind: simple faith."

While in Tibet during the summer of 1993 I encountered an example of Turrell's "artistic discovery," in the context of a traditional religious environment involving both faith and practice, which may further my case for vindicating the Desert Fathers. Within the Norbulinka, or Summer Palace of the Dalai Lama in Lhasa, there is a small courtyard where petitioners awaited entry into the reception room of the Dalai Lama. Lying down on the pavement of that courtyard, I was struck by effects of celestial vaulting and, I think, a kind of Ganzfeld response in which neural imagery flashed in my brain. Thus conditioned, when I visited the adjoining rooms, I experienced through the vast narratives of Tibetan Buddhism painted on the walls an explicit revelation of "other worlds," each well-landscaped and "ecologically flourishing."

A kind of paradisiacal format seemed to permeate these "visions" which more than described, but rather explicated a wide variety of alternative world occupancies and evolutionary and/or narrative scenarios held in the Vajrayana tradition. Here were "Desert Fathers" who maintained other worlds in forested and well-watered splendor, quite unlike the barren, tree depleted, rockiness of their present day environment. These "invisibilities" belied present realities, yet for the faithful existed fully on other dimensions—not as "paradises," but as environments of past, present (Shambhala) and future. In these examples, without becoming a Tibetan Buddhist, one can begin to experience an alternative reality in which spiritual perception (mythos) begins a dialectic with bios here and now. A primordial "paradise" assumes concrete dimensions in real time through visualization, meditation and mantra in a graphic, temporal, yet highly spiritual modality.

Thus, more practically speaking—the current Dalai Lama proposes an end to the ecocide and genocide being perpetrated by the Chinese in contemporary Tibet. He seeks a kind of "Peace Park" and natural preserve for flora and fauna and mineral earth and waters throughout his occupied country as a kind of eco-peace model for the entire planet. In his proposal, perceptions of Shambhala and other mythical eco-zones in "natural balance" (experienced in Tantric practice) are proposed for the here and now in Tibet today. The fact that the Chinese are not encouraging this "use" of Tibet does not alter the fact that the Tibetan Buddhist perspective carries within it a highly ecological and "natural" plan for coexistence in a kind of Gaia matrix.

These examples cannot "undo" the violent course of development so clearly and devastatingly outlined by Shepard in his schema explicated in *Nature and Madness*; but the mere possibility of such a dialogue in which the natural dialectic of bios and mythos is manifested may provide some modicum of hope for possibilities within the so-called larger, modern world in which so many of us find ourselves. In the work of Helena Norberg-Hodge we find a Western exponent of the transference of these archaic spiritual and practical perspectives to what she calls "ancient futures." Her Ladakh Project effects a challenging interface between late 20th century appropriate technologies, ancient practices and the forces of modernist development in Ladakh (a culture closely allied to Tibet).

Thus, if one had it to do all over again in those "Desert As a Place" course lectures, he would shift the ground from the Near Eastern deserts and the various manifestations of Hebrew and Christian Desert Fathers we find there to a counter-example half a world away in the Tibetan highlands, extending in their influence throughout significant parts of northwest China and Central Asia. That counter-example affords little evidence at present for succeeding in a reverence for land and a biophiliac's need for more natural developmental patterns, but the seeds for a spiritual ecology within which such might be the case is clearly more evident there than in our Western model so cogently analyzed and critiqued by Paul Shepard.

Notes

1. This is edited from my essay noted above which appears in the Pitzer College Catalogue "About the transformative effects of a winter trip to Arizona, meeting James Turrell, -9 degrees at Roden Crater and an aperture in Claremont," edited and published by students of the Turrell Independent Study Project under the direction of Professor Michael Woodcock.

Future Madness

John A. Livingston

NO DOUBT MOST IF NOT ALL contributors to this volume will have commented in some form at some time upon the global "development" idea. Having myself delivered a quite sufficient number of harangues on that subject, here I shall dwell instead upon two symbiotically intertwined concepts which serve to support the "development" imperative. The twin symbionts are (a) sustainability and (b) the future. Both are cultural and ideological artifacts, abstract social fabrications.

Let us begin with sustainability. What is it that is to be sustained? What is it that we are to maintain, prolong, nourish, and support indefinitely? If, as it appears to be, it is the advancement of the human monoculture, that is straightforward enough. Monocultures do two things—they homogenize and they simplify. Everyone knows that homogenization and simplification are destructive of natural communities. The human monoculture is death to Nature. That is the implicit end of the "development" of the biosphere. We are being instructed to sustain the explicit means.

As a naturalist I have always believed that the biosphere was already fully developed long before we began to homogenize and simplify. At any point in its career, the biosphere was and is fully developed. It is now, it was when some space junk blew away the dinosaurs, it was when the first crosssopterygians came ashore, and it was in the beginning. I have never understood the perceived necessity of the biosphere being "improved" in the image of a single species, especially one so grotesquely maladjusted, in an ecological sense at least, as our own. That is how I comprehend the advancement of the human monoculture. That is what we are being entreated to sustain.

But maybe that's not it. Maybe what is to be sustained is not business as usual, but rather some revised, scaled-down version of the human project. Maybe we should facelift the biosphere not in the image of the urban-industrial monster but rather in the image of the shepherds and the shepherdesses gambolling on the greensward in some New Age Arcadia.

That would be nice. There would be sheep, and there would be music, and there would be sunshine and puffy cumulus clouds—just little ones, mind you—

and all would be in harmony with Nature. Trouble is, any really heavy-duty gambolling eats up a hell of a lot of greensward, and what with the sheep and all, there wouldn't be room on the planet for six thousand million frolickers. Let alone any gardens or vineyards. Or anything else, for that matter. So I reckon it's not that either.

Maybe what is to be sustained is renewable resources. The term "renewable resource" has itself gained full-blown oxymoronic status. Codfish, great whales, old growth forests, tropical soils, fresh water aquifers—even frogs and songbirds—all used to be called renewable resources. We need not invest much cerebration on that one. Unless there's some way to cash in on Great Lakes zebra mussels, which renew themselves with quite astonishing fruitfulness—and without the help of any resource management expertise whatever.

I suppose the ultimate beneficiary of sustainability—at least rhetorically—would have to be life process itself. We need not worry about that one either. Life process will go on—with us, or without us. That, by the way, is to be construed as a positive statement.

But still we have the sustainability imperative. In whose interest are we to sustain whatever it is that we are to sustain? Nature's interest? Not bloody likely. It must be the interest of the industrialized nations. Or is it that of the would-be industrialized nations? Or of simple human fecundity?

Or, maybe it's the interest of an even higher purpose. I think we're getting warmer. Maybe what is to be sustained is not a concrete tangible but an idea. An abstraction of some kind. Maybe what we are being asked to maintain is the very concept of the future.

We thrash about in a maelstrom of ironies. One of the more crashing is our collective pride (hubris, more like) in our abstraction. Other animals regularly abstract in their daily lives (problem-solving may be universal in vertebrates at least, and there is plenty of evidence from octupi, among others), but the human does very little else. Abstract reasoning is the human specialty. But specialization, like the free lunch, always has its cost.

Along our improbable evolutionary way, abstraction seems to have become increasingly dominant relative to the rest of our "normal" animal gifts. So important did it become that at some point a qualitative Rubicon was crossed, and we became no longer the masters and possessors of abstraction, but its creatures.

We moved into a state of total dependence and servitude, most especially to instrumental (technical) reasoning, and in so doing lost our participatory relatedness to and identification with the wider life community. No longer able to perceive the rest of the multitudinous biosphere in any way other than the abstract and utilitarian, we had become ecological loose cannons, dangerous misfits in what had been theretofore an ineffably good, perfectly functioning and integrated community of being. We found ourselves alone.

We rationalized our loneliness as uniqueness. That was pretty wide of the mark. All species, all individuals are unique. So we rationalized our loneliness as ordained separateness, prefigured noblesse. God said that there should be apartheid on Earth. We liked that, so we wrote it down, and codified it. So went the reasoning—or, as I prefer, the rationalization.

There is a monumental irony in our specialization, and that is in the abstraction of time. More precisely, the abstraction of the future. Obviously, all technical or instrumental thinking and tinkering, and all "progress" orientation requires the prior assumption not only of perfectibility but also of goals, and the assumption of goals in turn necessitates the prior abstraction of the future. Forward-thinking by most well-adjusted animals appears to be confined to the duration of the task at hand— the strategy of the hunt, the best approach to the immediate challenge, the weighing of present options in the here and now.

Non-human beings and a shrinking number of human societies have always lived in the joyous fullness of the present. Our society has utter contempt for their apparent inability or unwillingness to be victimized by morbid obsession with what may or may not obtain tomorrow.

Our future-orientation is virtually total, certainly at least in Western cultures. Our self-identities as individuals and societies and civilizations—maybe even as a species—appear to be cast in terms not of what we have been, not of what we are, but of what we shall become. But remember that only humans become. Non-human beings may be seen as evidence of the creativity of natural selection, but humankind must be seen differently. Our species must be seen as the eventual and final culmination and realization of life's purpose. The future of planet Earth is the human future.

A difficulty arises in the fact that by definition the future never comes. It follows that in practice we can never become. Our self-identity can never be found. Having long since rejected our self-identity as Nature, we chose to put all our chips on the future, whose margin, like that of Tennyson's struggling Ulysses, fades forever and forever as we move. No matter; the priests and the secular gurus, the technocrats and the humanists generally, keep flogging our progress toward future self-realization.

But if we identify solely with the future, then surely we never grow up. This is, of course, the common condition of all domesticated mammals, among whose number I have recently argued the human species must be included.[1] Paul Shepard has shown that Western society itself suffers from arrested development.[2] (That was in 1982; in the 1990s perhaps we must say that Western society is developmentally challenged. Which indeed it is, in more ways than one.) As individuals, societies, and civilizations, we are forever becoming, forever questing for mature adulthood.

It follows that, in the advancement of our quest, whatever must be done, must be done. If it means the simplification and homogenization of natural communities and processes, so be it. If it means the extirpation of whole floras and faunas—without replacement—then that is in the nature of things.

The human enterprise is necessary. The human future is imperative. The subsidy Earth and earthly nonhuman beings are required to pay for human immortality is of course factored out, not in. That is an externality—an incalculable, as the technocrats like to say. Regrettable, perhaps, but necessary and inescapable.

My final question concerns how the future, however imagined, is to be sustained. The idea of the future we sustain by mantra, litany, catechism, ideology, casuistry, and a good shot of group self-mesmerism, to say nothing of human chauvinist narcissism. The realization of the future—the necessary advancement of the human monoculture—is to be sustained by Nature.

The clear assumption is that Nature owes us. It is Nature's appointed task—its reason for being—to maintain and nourish the human project. Nature was provided to serve the Chosen Species. That is the received cultural and historical wisdom that sustains the ideology of the necessary primacy, among all things, of the human enterprise.

I do not know whether all forms of madness need inevitably be tragic. But this variety has all the earmarks.[3]

Notes

1. John A. Livingston, *Rogue Primate* (Toronto: Key Porter Books, 1994).
2. Paul Shepard, *Nature and Madness* (San Francisco: Sierra Club Books, 1982).
3. Livingston, John A. "Les implications ecologiques du caractère impératif de l'immortalité de l'homme" *Frontieres* vol.5 nu.2 (Automne 1992) and "Nature for the Sake of Nature" in Monte Hummel, ed., *Endangered Spaces* (Toronto: Key Porter Books, 1989).

Warrior's Wound, Woman's Woe: The Fall from Grace

James A. Mischke

> The spear's interpenetration of the body and the flesh as the source of all new life are the iconography of venery—at once the pursuit of love and game. . . . The relationship between them does not depend on intellectual symbolism. Both have to do with the most profound of life's passions, the demonic moment of the kill and of orgasm. These two powerful expressions are related. Both lead to life and to death, which, like maleness and femaleness, represent a fundamental polarity.[1]

THE GODDESS WHO BEARS THE BURDEN of balance between life and death, thus aiding at once hunter and hunted, is known to the Athapascan (Navajo language group) speaking hunters as the Mistress of Animals. In his consideration of this most pivotal of mythical characters, Karl Luckert draws an analogy with the Greek goddess Diana, "still in her youth, still out in the wilds and surrounded by genuine hunters. The southern Athapascan Diana is a sister of the hunters, sometimes even the wife of a hunter. She leads a life in twilight between the family of hunters and the family of deer, between killers and victims. She is the female servant of the hunter's household, all the while she is denied the full participation in hunter life."[2] We begin to glimpse the theme of domination and lack of respected status experienced by the goddess who has previously provided the hunter with the cosmological position of guiltless survivor and successful predator as well.

Paul Shepard speaks of the mirroring in nature of the thanatotic and erotic instincts as may be witnessed in the reciprocity between the hunter and his woman. This "posturing," he maintains, on the part of the female, which transforms rage into lust has indeed been the secret of survival of many if not most species. Even among reptiles and birds,

> the first steps in courtship are attack and deflection. With a sinuous movement, the female lifts her head, flashes a color, gives a call, or produces an odor—and the attack is instantly transformed into part of a dance, which will elicit a new response from the female. A courtship has begun with a beautiful set of signals initiated in the style—for a moment—of aggression.[3]

Shepard goes on to extrapolate from the feathered to the primates and so on to Homo sapiens: "The woman's special morphology is then bought into play as eros inciting signals, emphasized by modified submission postures, such as that called 'presentation' by primatologists."[4] The use of feminine morphology in the redirection of the "venereal aggression" of her hunter mate—from the kill to the orgasm, in the draining off of masculine "in-tension-ality" in its excess—gives us an example of culture's identity and function as a superstratum of instinct, acting as a conduit for the conduction of instinctual thrusts into the societal realm in forms that may be mutually read and reacted to. "For the same purpose are those culturally refined, eroticizing, cryptandric features...such as jewelry, bound feet, bustles, hair arrangements, perfumes, and tattooing" utilized in such uniquely human readaptation.[5] The hunter, consciously or unconsciously, understood the relationship between love and death, the two great mysteries of existence. The thanatotic and erotic were linked through a set of symbols which rendered for the hunting male an experience of the erotic as a form of the killing and the devouring. Both imply an enactment of sacred relatedness upon which totemic exogamy is based. Shepard reflects as to the eating of the flesh of other creatures by the human carnivore. "Symbolically, it is a re-enactment of the emergence of man from herbivorous ancestors and the dark of ignorance, as love between man and woman is a re-enactment of the discovery of love as a continuing bond, a passion engendering compassion and affection, prefiguring the feeling for and affirmation of other species and the world."[6]

What the hunter developed, however, the warrior lost. Warrior, as the offspring of man's economic and spiritual perversion, displays though his sexuality his inability to participate in sacrament. In contrast to the hunter, the warrior knows little of sublimation of thanatotic impulse into passion for one's mate, and in his own patriarchal fear of death, hates and fears the bearer of the complementary pole of the erotic, the woman. Thus, the warrior desecrates his relationship to his Universe, his mate, and himself in the attempt to outwit rather than to celebrate nature; the dynamic equilibrium provided by the totemic, which linked instinct and culture, was violated. Erotic love, as an integral part of human experience, became split off from those symbols which activated it and explained its meaning. Its bearer, woman, while instinctually behaving properly in displaying behaviors of attack deflection and general primate posturing, came to be regard as "temptress." As such, her ability to deflect the warrior's destructive thanatos became limited; to the degree that courtship deflection became "culturally normative," and the assault of the former hunt came to be partially visited upon the sexual partner.

In woman's degradation is society rendered imbalanced; the former hunter has but one remaining outlet for the thanatotic impulse: war. Such is the plight of the pastoralist. He is at once creator and victim of the cruel warrior Father God, whose accusatory finger was first pointed at Eve. In the eroslessness of the repressive Yahwehs and Allahs lies the preclusion of the warrior's anima development. Within the implosion of "the bipolarity of the passionate" lies the source of sexual guilt of "civilized man"—the quandary of the confusion born of inability to make contact with the transformative symbolic which facilitates the transfixing of fury by passion.

Life and death constantly interface and it is most likely the instinctual thrust of managing such balance which preserves life. It was the basis of much of the culture of the hunter, certainly of the Athapascan. The enactment of the themes of the prehuman flux (the belief on the part of hunting peoples that transformation between human and animal is readily attainable) is applicable to the genders and the functions of both poles of thanatos and eros. The male clothes himself in the robes of the deer in order that he may hunt. What of the female who has ready access to similar attire while in her husband's absence? If her mate uses such tools of identity to the ends of his thanatotic purposes, might she not likewise utilize the skins for the expression of her erotic nature: "From the moment on when hunters dressed themselves and became deer for the purpose of killing, the equal rights of women to become does for the purpose of procreation had to be acknowledged."[7] The Athapascan tell the tales of the sister of the hunter who had eloped with the buck, to be "saved" only through the brother stunning her with his arrow and forcefully stripping away the hide to restore her to her human state of nakedness, thus overcoming the influence of the will of the animal gods that the human render up its identity in the service of the atavistic entropic thrust of the natural law of prehuman flux. The animal mistress is not to be allowed to return to the natural, but rather, due to masculine insistence upon the continuing stream of flow toward consciousness, is to be carried along with the current.

Through this allegory we are able to comprehend the agony of tension born by the hunter society in living in the force field between death and life: predation and procreation as thematic content within the containing stricture of transformation of identity between human and beast! And it is art of this thematic matrix of identity confusion that the final blow is dealt to the Navajo hunters blessed relationship with the Mistress of Animals. It is the archetypal trickster/hunter, this time in the form of the shaman, who strikes the goddess with the irretractable blow of desecration.

In the world where the verb is the predominant paradigm, how things work on the outside is logically translated through introjection to an understanding as to how things work within oneself. In a world where the natural and the zoological are the reality of concern, the interrelationship of other species becomes the ground for relationship between humans. The pattern of interaction between complementary entities is the issue to be addressed. Each form of life must simultaneously interact with others while at the same time maintain its own separate identity, a theoretical description of the basis of systems theory, par excellence! The external thanatotic dance of the hunter concerns itself with which prey will be devoured by which predators under what conditions. In the sense that woman is the guardian of societal relationships, her concern is that of the legitimate discharge of erotic energy. The latter is patterned after the former. The question as to "How does humanity fit into the web of life?" is answered by masculine cynegetic predation extra-systemically; it is answered by the goddess through the structure of totemically defined exogamic relationships. Shepard writes: "The relationship between clans is defined by selected details of the relationship between their totemic animals according to a myth."[8]

Thus, the concept of the interdependency of the many species acts in totemic society as the symbolic basis for human social organization. Herein we not only witness the basis of what forms of relationship preclude or legitimize marriage, we also, though the function of clan structure, come to understand the subtle metaphor of human relations under a conceptual system which seems to have been lost on men who "work" under horticultural and industrial "reality": men do not annihilate each other (this even though the various species of totem animals representing interacting clans may well prey upon each other). Over the aeons of time necessary for development, totemic society has demonstrated its ability to utilize metaphor toward survival ends.

While the killing of the beast is the mainstay of human association and its supportive rites and rituals, the young shaman hunter has gone too far, for this time the victims struck down are none other than the "twilight fawns," those blood relatives dwelling in the mist between societally agreed upon actuality and the archetypal realm of the animal gods. The act of infanticide has in fact bought down the children of the Mistress of Animals. Familial boundaries are violated; family love, loyalty and trust are ruptured. The poles of life and death have been brought too close. The structure of relationship alluded to by the universal truth of the totemic has imploded. The prey no longer dies through sacrament but through "murder." Time, the "actuality" of history as replacement for the metaphor of myth, and its precocious child, consciousness, are on their way.

The archetypal trickster/hunter reappears as he brings down the twilight fawns, thus vanquishing the balancing feminine element back to the unconscious, rendering male thanatotic energy its sway in the conscious world. Thus we view what is perhaps the earliest oral tradition recording the sacrifice of feminine psychological energy in the growth of consciousness. The hunting grounds are rendered imbalanced; predator begins to harvest more than a sustainable yield. The death of the prey supersedes that of the eros force; hunting has sounded its own death knell. The goddess does not reappear until the advent of horticulture when a reenactment of the release of the Animal Master's pets (which became the original prey of the hunt) by the Promethean hunter is achieved in Eden: Adam filches the fruit from Yahweh's garden. Within this scenario of the garden, the balancing female principle, the wife of the "master of the house," is straightaway struck down and rendered ineffective except as a carrier for projected guilt between abusive father and Oedipal son.

Upon attaining birth, the Navajo were, with their goddess, struck down by that pastoralist thrust, which long before had overwhelmed Abraham's foremothers; thus the Dine, like their anteceding Hebrew brothers, fell into the hands of an erosless and irate god of war.[9] The Navajo mythic version depicts the maternal principle of the horticultural as a viable element until the advent of the Warrior twins. Soon after, Changing Woman is transformed into a "Warrior Woman" during the historical time that Spanish livestock is rapidly falling into Navajo hands. Navajo or Judaic, the common ground reflects the singular theme that consciousness and guilt flourish only in each other's presence, yet always at the cost of death's dominion over life, that realm of which the principle of feminine consciousness is patroness.

Humanity no longer lives in the realm of the sacred, but rather in the absence of the erotic; humanity falls again and again in and through the impalement of the shaman's arrow. Humanity has set on high the divine animal mediatrix. No brother's arrow can retrieve her now. Having withdrawn from human society, the goddess has flown to heaven where she is now to be reached only through formal religious activity. Gender split is immortalized. The worlds of beast and "man" are irreconcilably cast asunder. The distance between hunter and hunted now receives explanation and justification. The goddess' original work as matroness of distance between man and beast is accomplished, yet circumvented.

Symbolic meaning had gone awry. A twisting arose within the shared system of symbolic meaning of the West as to the essence of the beast. Sin and evil became that which was "animalistic"; what was once freely and innocently erotic, the interplay of life and death, now lay in the ashes of collapsed polarity once manifest between eros and its opposing thanatos. It has come to be known as duality:

> Persian duality helped the Christians transform all ambiguity into opposition instead of metaphor. The efforts of religious energy to reanimate the cosmos merely had the effect thereafter of creating sides, populated with demons and devils on one hand and saints and angels on the other. Augustine saw the desert itself as a war. Animals represented evil. The anchorites, like the modern fundamentalist, were eager to fight devils "in the freedom of the vast wilderness," according to John Cassian...in order "to find that life which can be compared to the bliss of the angels." The hermit-animal traditions, involving Jerome and John the Baptist, meant subduing that ferocity which was supposed to have resulted from the Fall.[10]

That which was in actuality the pressing thanatotic hunting instinct in us all, clawing for expression, that which had been our tool of survival even before our humanity, we have come to despise—all as a function of our cultural misinterpretations arising out of the perversions so endemic to the practice of animal husbandry.

The objectification of the world and its multiplicity of life forms, the loss of its enchantment, has been brought about by "Man." This final, ultimate thrust of "the Fall," away from universal consciousness into the perception of a dead universe, was reacted to by Blake in 1802 as the result of a vision: "May God keep us from single vision and Newton's sleep." The erotic in the midst of the groaning masculine human thanatotic cannot longer be born within the "bipolar disorder" of human culture. The shamanic trickster has, by release of the arrow's shaft, brought to conclusion "the creative event," thereby separating heaven and earth; in trade for the release of and from universal tension, thanatos is on the march.

Notes

1. Paul Shepard, *The Tender Carnivore and the Sacred Game* (New York: Charles Scribner and Sons, 1973), 170.
2. Karl Luckert, *The Navajo Hunter Tradition* (Tucson: University of Arizona Press, 1975), 140.
3. Shepard, *Tender Carnivore,* 172.
4. Ibid.
5. Ibid., 172-73.
6. Ibid., 173-74.
7. Luckert, *Hunter,* 141.
8. Paul Shepard, *Thinking Animals* (New York: The Viking Press, 1978), 129.
9. It was this god of the two brothers, Cain and Abel, to whom the appeal was made to accept the "fruits" of their respective labors. Cain, the tiller of the fields, was rejected in favor of his pastoralist brother, Abel, who was able to provide sacrificial substance in the form of the flesh of the beast once hunted and taken through sacrament in the wild.
10. Paul Shepard, *Nature and Madness* (San Francisco: Sierra Club Books, 1982), 61.

PART III

ESSAYS IN HONOR OF PAUL SHEPARD

The Fate of Jefferson's Farmer

Douglas L. Wilson

(Reprinted with permission of North Dakota Quarterly and the author.)

ONE OF THE GREAT WINDFALLS of my academic career was starting out on the same college faculty as Paul Shepard. When I landed at Knox College in 1961, it soon became evident that a gentle, soft-spoken biologist was not only regarded as the leading scholar on the faculty, but that his ideas had made an noticeable impact on his colleagues. They were all aware, for example, of the prime significance of ecology, a study with, at that time, no standing or general recognition whatsoever in the public at large. His scientific interests reached into so many other disciplines—art, history, philosophy, anthropology—that he was, for a liberal arts faculty, an ideal colleague. As my own intellectual and research interests developed, Paul Shepard's wide-ranging model, which ignored traditional disciplinary boundaries, proved an inspiration, and touching base with him from time to time became a source of stimulation and encouragement. By the time his book *The Tender Carnivore and the Sacred Game* was published, I had realized a long-standing dream and had acquired a small farm, which is where I first read his chapter "Farming: An Ecological Disease." Since then Paul and I have been arguing back and forth about farming where I play the role of championing it as a cultural ideal and he regularly enters his reasoned ecological demurrers. As his readers are always aware, Paul is unlike anybody else, but I have been struck over the years with certain points of affinity between him and the subject of much of my writing Thomas Jefferson. Like Jefferson, Paul is one of those troubled by the fact that Americans are generally more interested in turning a profit than in living well, particularly in matters that affect the physical environment, (This is well expressed in the sardonic title of a book he once half-jokingly proposed to write: Let's Pave Iowa.) Paul also shares Jefferson's doubts about cities as viable human habitats, a point of view that may finally be coming around for serious consideration. So I propose to pay tribute to Paul by dedicating to him the following essay that deals with Jefferson, farming in America, and even touches on the state of things in Iowa.

I

During the Revolutionary War, a French request for information on the state of Virginia found its way to Thomas Jefferson, who approached the task with characteristic energy and originality and ended by producing a substantial book, *Notes on the State of Virginia*. In revising these notes for publication, he added to a section on manufacturing a passage in praise of farmers and farming which is among his most famous utterances: "Those who labour in the earth are the chosen people of God, if ever he had a chosen people, whose breasts he has made his peculiar deposit of substantial and genuine virtue. It is the focus in which he keeps alive that sacred fire, which otherwise might escape from the face of the earth."[1] From this resounding compliment, American agriculture has never quite recovered.

Jefferson was not here simply playing the politician's role of buttering up the populace, which was mostly farmers. In fact, he had recently retired from the governorship of Virginia under a cloud of undeserved criticism and had made up his mind to take permanent leave of politics. His paean to agriculture invoked a very old idea that was deeply rooted in the classical tradition, a body of writing and sentiment with which he was intimately familiar. It held that country dwellers, and particularly those who tilled the earth, were more fortunately occupied and more virtuous than their urban counterparts. The eighteenth-century Enlightenment had taken this ancient notion very much to heart, and it fell to Jefferson, its foremost American spokesman, to pronounce the classic formulation. "Corruption of morals in the mass of cultivators," he wrote,

> is a phaenomenon of which no age nor nation has furnished an example. It is the mark set on those, who not looking up to heaven, to their own soil and industry, as does the husbandman, for their subsistance, depend for it on the casualties and caprice of customers. Dependence begets subservience and venality, suffocates the germ of virtue, and prepares fit tools for the designs of ambition.[2]

Jefferson, of course, had more in mind than simply rehearsing the familiar theme of Virgil's *Georgics*, which was undoubtedly one of his sources. Rather, he was recasting a classical ideal in a form readily applicable to the political and economic conditions of Revolutionary America. There were few large cities in the colonies (none at all in Virginia), and the vast majority of the citizens were farmers who subsisted for the most part on the products of their own labor and exercised a large degree of independence. This accounted for the obvious advantages, as Jefferson believed, enjoyed by America over Europe, where most of the land is "locked up against the cultivator" and manufacturing "must therefore be resorted to of necessity, not of choice."[3] What a mistake it would be, he reasoned, if Americans were to forsake the conspicuous advantages of an agricultural society, which they already possessed in full measure, by trying to imitate the Europeans.

> While we have land to labour then, let us never wish to see our citizens

> occupied at a work-bench, or twirling a distaff. Carpenters, masons, smiths, are wanting in husbandry: but, for the general operations of manufacture, let our work-shops remain in Europe. It is better to carry provisions and materials to workmen there, than bring them to the provisions and materials, and with them their manners and principles. The loss by the transportation of commodities across the Atlantic will be made up in happiness and permanence of government. The mobs of great cities add just so much to the support of pure government, as sores do to the strength of the human body.[4]

This is the essence of Jefferson's theory. But he was quick to point out to a correspondent shortly after he had published these words that it was "theory only, and a theory which the servants of America are not at liberty to follow."[5] The reason for this was partly due to the unwillingness of Great Britain and other trading nations to acquiesce in so equitable a system, but the main obstacle was actually to be found in the inclinations of the American people themselves. As Jefferson had conceded in the *Notes on Virginia*, "the actual habits of our countrymen attach them to commerce."[6] This important qualification has been conveniently overlooked by Jefferson's detractors, from his day to our own, and his vision of an agrarian society has more often than not been set down as an Arcadian fantasy, inspired by the misty-eyed poets of pastoral.

But Jefferson was too pragmatic and, for all his idealism, too much of a realist for this. He may have been inspired by the verses of Virgil, but it was hardly the lovelorn swains of the *Ecologues* that put him in mind of the American farmer. It was rather the hero of the *Georgics*, the sturdy aqricola who, participating in the natural rhythms of the seasons and presiding over his small domain with the prerogatives and responsibilities of a king, attained happiness through hard physical labor and self-reliance. Nor was the notion that American society at the time of the Revolution approached the agrarian ideal depicted by Virgil solely Jefferson's observation; it was, for example, independently and forcefully confirmed in Crevecoeur's *Letters from an American Farmer*, which was written at about the same time as the *Notes on Virginia*.

What most threatened this fortunate state of affairs, in Jefferson's view, were the propensities of the American character itself. "Our people have a decided taste for navigation & commerce," he admitted sadly; "They take this from their mother country: & their servants are in duty bound to calculate all their measures on this datum."[7] This is why, as one of its servants, he was unable to indulge his own theory, as he put it, and instead labored for five years in Paris after the Revolution to secure treaties and commercial trade agreements that would insure safe navigation and favorable European markets for American goods. Throughout his life he would acknowledge the practical necessity for native manufacturing and foreign trade, but he never gave up his conviction that farmers made the best citizens and that agriculture presented the greatest opportunities for a satisfying way of life.

The canker in the rose, as Jefferson saw clearly, was that Americans were all too interested in turning a profit, perhaps even more interested in that than in living well. In fact, it might be argued that Americans in general have never been able to understand the latter except in terms of the former. Material prosperity has, from the beginning, been regarded as a providential sign that one's efforts were worthy, and Americans have always been looking for signs. For Jefferson, the ultimate source of satisfaction was freedom and self-reliance, something that only a tyrant could threaten, but the aspirations of his countrymen, from his day to ours, have usually been focussed on things more subject to moths and rust.

Jefferson rightly supposed that the availability of cheap land would continue, as long as it lasted, to attract Americans to farming. In a world where the chief access to wealth was land, the landless flocked to the frontier to take possession of their allotted portions in spite of bitter hardships and deprivations. Clearing the timber, breaking the prairie, and putting up improvements indeed made them farmers, but it also usually made them debtors, which in turn only helped to push them into a role they usually preferred, that of entrepreneur. Taking whatever profit might have been created by one's labors on the newly acquired land and moving West to try again became much more common than sending down roots and cultivating the good life. Frederick Jackson Turner has called attention to the values and mores that the frontier experience seems to have produced, but there can be no doubt about the fact that it did not produce a nation of happy farmers. By the time the free public lands were exhausted and the frontier was closed at the end of the nineteenth century, American agriculture had become a breeding ground for discontentment, and farmers were rapidly becoming a race of tenants who consistently failed to share in their country's expanding prosperity.

II

Whatever happened to Jefferson's agrarian ideal—a nation of self-reliant freeholders and farmers? In one sense, it died aborning. Though it seemed at first to fit the conditions of a people who were farmers by default, it demonstrated little appeal as a practical goal for a nation of restless achievers. Jefferson's great fear was too soon realized, as manufacturing took hold with a vengeance and the cities began to dominate the cultural and physical landscape. In a memorable sentence that sums up our history, Richard Hofstadter wrote: "The United States was born in the country and has moved to the city."[8]

And yet, in an important sense, Jefferson's agrarian ideal has survived. It not only lived on in the nostalgic memories of millions of Americans who grew up on farms and left them, it has survived tenaciously in the American imagination, and even at a time when significant contact with rural life is restricted to a small and diminishing minority of the population, it still exercises an enormous appeal. American writers, politicians, cultural confidence men of all descriptions—"all of us in the trade," as Mark Twain put it—have always understood instinctively the power of what Hofstadter called "the Agrarian Myth." This may be described as an emotional attachment to rural life and values, which is linked to a fanciful vision of America's rural past. The persistence of the Agrarian Myth is, of course, nothing less than a cultural paradox in America, whose citizens have, in the critical choices of their own behavior, historically opted overwhelmingly for urban life and values.

The most notable champion of rural life in recent years has been the poet Wendell Berry, who has written eloquently on the plight of the small, traditional farmer and the relentlessness with which he is threatened by the forces of industrialized agriculture. Berry is no confidence man; he is a first class writer who has portrayed the nobility and satisfactions of true husbandry and has acted them out in his own experience. His writings are implicitly a testament to the agrarian ideal that Jefferson enunciated, and he has urged anew its viability and importance in a culture that has lost touch with its roots and seems bent on its own destruction.

Berry's writings on American agriculture are so brilliant in their literary craftsmanship and so noble in their intentions that it seems almost a sacrilege to find fault with them. On the level of morality, they are perhaps as forceful and effective as any ever offered on the subject. When he makes the case, for example, for the nurturer over the exploiter in the opening pages of *The Unsettling of America,* his comparisons are compelling:

> The standard of the exploiter is efficiency; the standard of the nurturer is care. The exploiter's goal is money, profit; the nurturer's goal is health—his land's health, his own, his family's, his community's, his country's. Whereas the exploiter asks of a piece of land only how much and how quickly it can be made to produce, the nurturer asks a question that is much more complex and difficult: What is its carrying capacity? (That is: How much can be taken from it without diminishing it? What can it produce dependably for an indefinite time?) The exploiter wishes to earn as much as possible by as little work as possible; the nurturer expects, certainly, to have a decent living from his work, but his characteristic wish is to work as well as possible.[9]

But when it comes to acknowledging which one of these historically characterizes the traditional American farmer, Berry lapses into the Agrarian Myth.

He seems to think and would have us believe that exploitive farming is something that is not characteristic of traditional American agriculture and is largely the product of a conspiracy among the government bureaucrats, agricultural schools, and the moguls of agri-business. That American farmers have been all along generally more interested in short-term profits than permanent productiveness is implicitly denied, though this is the clear and unequivocal verdict of our agricultural history. George M. Fredrickson, in a recent review in the *New York Review of Books,* sums the matter up this way: "Commercialized agriculture—production for markets beyond the local exchange community, the view of land as a commodity rather than as a permanent source of family subsistence, and the willingness to move to new lands with greater fertility or access to markets—has been seen as the norm since the late eighteenth century, if not earlier." In spite of the efforts of some historians to demonstrate the persistence of a more Jeffersonian agriculture, with its emphasis less on markets than on freedom and self-reliance, Fredrickson allows that such efforts have failed to dislodge the prevailing view "that most rural Americans of the late eighteenth and nineteenth centuries were expectant capitalists who valued the prosperity promised by industrialization and the growth of market more than they feared the loss of independence and self-sufficiency."[10]

This is not to deny that there have always been American farmers who have regarded farming not so much as a business but as a profoundly rewarding way of life, who are less concerned with the marketplace than with their moral responsibilities as members of a community and husbanders of the soil. And Wendell Berry is undoubtedly their finest spokesman. Berry does not present himself, of course, as a hard-headed agricultural historian; he writes as a poet, a teacher, a farmer, and a Kentuckian. He is wonderfully eloquent and incisive on many aspects of rural life, especially on the values and virtues of the small farm, and he can be just as effective as the scourge of agri-business or the bureaucratic mentality. But he is undoubtedly soft on farmers, slow to blame and eager to explain and exculpate them as a class. Whether or not this should be considered a fault, when coupled with his historical shortcomings, it seriously undermines his attempts to explain what has happened to American agriculture.

Consider his argument that the shift to bigness in agriculture that is driving farmers off the farm is "not primarily the work of farmers, though it has burgeoned on their weaknesses." Fair enough, if we include among their weaknesses their notorious weakness for short-term profits. This development, he goes on, is primarily "the work of the institution of agriculture: the university experts, the bureaucrats, and the 'agribusinessmen,' who have promoted so- called efficiency at the expense of community (and of real expense), and quantity at the expense of quality." But note what happens next: "And along with the rest of society, the established agriculture has shifted its emphasis, and its interests, from quality to quantity, having failed to see that in the long run the two ideas are inseparable."[11] (42). Berry's point about the relatedness of quality and quantity is so well taken that it is easy to lose sight of the fact that we are being presented with an historical argument: American farmers (scarcely recognizable here as "the established agriculture") used to be more concerned with quality but have been induced to shift their emphasis. While his point may well apply in certain historical situations, it obviously does not describe the case of American farmers generally, who, as expectant capitalists trying to improve their position in the marketplace, have been interested in quantity over quality all along. In fact, our seeing them in this perspective, as agricultural history forces us to do, causes Berry's whole conspiratorial theory about the victimization of American farmers to unravel. It becomes yet another eloquent chapter of the Agrarian Myth.

III

Berry is undoubtedly right about the fact that the emergence of an agricultural establishment in government and industry has vastly complicated the world of the American farmer. If it was Jefferson who famously scoffed at the spectacle of the government telling the farmer what to plant, it was also Jefferson who promoted the idea of agricultural education. As Berry understands so well, in our time the second seems inevitably connected to the first. Develop agriculture into a teachable science and you establish, willy nilly, an approved way of doing things (which, unfortunately for the farmer, is constantly changing). Farmers who refuse to go along with these enlightened methods can reasonably be categorized as uncooperative and (what is worse) unprogressive and therefore denied government welfare, which, as farmers gallingly admit, is what the modern farm programs have become. While it is folly to deny that agricultural education and government programs were fostered and are perpetuated to benefit farmers, the results have certainly been ambiguous, to say the least. Setting out to make the traditional American farmer more efficient and productive, they have succeeded so well as to make him virtually unnecessary. That one farmer can now raise what it used to take five to raise is fortunate for him, but the other four have thereby been made redundant. The question that Berry and others have raised is to what extent have farmers been railroaded into redundancy by bureaucrats and profiteers.

For what is wrong with the approach of government agencies, the wry wit of wildlife ecologist Aldo Leopold seems more apt than much of Berry's conspiratorial theory. In *A Sand County Almanac*, he writes: "The State College tells farmers that Chinese elms do not clog screens, and hence are preferable to cottonwoods. It likewise pontificates on cherry preserves, Bang's disease, hybrid corn, and beautifying the farm home. The only thing it does not know about farms is where they came from. Its job is to make Illinois safe for soybeans." Adding up the ecological price of what the agencies prescribe for profitable farming, Leopold pointedly asks "Just who is solvent? For how long?"[12]

The transformation of American agriculture has a long history, but until the second half of this century it retained some of the characteristics of Jefferson's agrarian ideal. Most of the farms were still small and family operations. Some significant portion of the family's food was raised and processed on the farm—milk and milk products, eggs, poultry, fresh and home-canned vegetables, pork, and beef—and the unneeded excess traded for groceries and other goods. Farmers and their families were, by virtue of their physical isolation, largely deprived of the amenities afforded by town and city life, and thus led a more primitive existence. But this changed dramatically following the second World War, which, with the Marshall Plan, brought a burst of prosperity to American farmers. Vastly improved transportation and communications now gave farmers ready access to urban amenities; things such as electrification, indoor plumbing, reliable telephone service, and television became commonplace; and farmers were finally able to participate to some extent in the general material prosperity that was giving rise to the affluent society.

These developments of the recent past significantly changed the lifestyles and heightened the expectations of those remaining on the farms and might be said to have brought about the final undoing of Jefferson's dream.[13] Old ways were gladly exchanged for new ones, and these invariably entailed greater expense, increasing the need for disposable income and credit. Given a choice, farmers were no more willing than other Americans to forego membership in the affluent society, where the good life was based more than ever on consumption, for which the first requirement was something the small farms had never been able to produce in quantity—money. The historical movement that brought Americans from the farm to the city was all that Jefferson had originally dreaded and feared: forsaking the virtuous rural society for the corruptions of the cities; giving up the benefits of working with the generative force of nature for demeaning taskwork; and trading the independence of the self-sustained farmer for total dependence of the wage earner. Along the way, the farm had ironically come to represent the enslaving life of drudgery, and the job in town, deliverance and freedom.

As Wendell Berry and others have shown, it is possible in our time for a small landholder to support a family through non-exploitive farming, and some, such as the Amish, have made a conspicuous success of it. It is still possible, just as it has been throughout our history, to make much of what one needs for subsistence, to raise most of one's food, and to live on a short cash margin. This is, admittedly, no easy feat; it requires hard work, discipline, and determination, but these are qualities that American have never lacked. The real difficulty is that it also seems to require the belief that a simple life of laboring in the earth is the most worthy and rewarding, something that most Americans can only affirm imaginatively, through the Agrarian Myth.

IV

But even though their numbers are vastly smaller and their capital investments immensely larger, farmers remain on the land. The plight of the current generation of American farmers, unmercifully squeezed by mounting indebtedness and falling prices and land values, brings to mind an illustrious example—the agricultural and financial career of Thomas Jefferson himself. Jefferson was not, of course, a small freeholder, the class he singled out as "the most precious part of a state."[14] He was, through inheritance and marriage, a large landholder, who farmed thousands of acres with slave labor, a system he deplored as morally wrong and sought to change. Moreover, the prevailing agriculture of Jefferson's day, particularly the method of raising the all important cash crop of tobacco, was ruinous to the land, something Jefferson also deplored. "It is a culture productive of infinite wretchedness," he wrote in the *Notes on Virginia*. "Those employed in it are in a continued state of exertion beyond the powers of nature to support. Little food of any kind is raised by them; so that the men and animals on these farms are badly fed, and the earth is rapidly impoverished."[15]

The solution he proposed was the cultivation of wheat. "Besides loathing the earth with herbage, and preserving its fertility," he wrote in his *Notes on Virginia,* "it feeds the labourers plentifully, requires from them only a moderate toil, except in the season of harvest, raises great numbers of animals for food and service, and diffuses plenty and happiness among the whole."[16] In addition, his extensive study of agricultural science led him to believe that crop rotation and the application of natural fertilizers could sustain the soil in a constant state of production, so that the disastrous exploitation of the land for short-term profits, such as prevailed in producing tobacco and Indian corn, could be avoided. But Jefferson's obligations as an office-holder kept him away from his farms for long periods of time, and he was unable to put his ideas into practice until he retired from public service in 1793 at the age of fifty.

For three years he labored diligently to repair the "barbarous state" of his lands and to put his farming operations into conformity with his theories, but his success was limited. He had difficulty getting his rotation scheme established, particularly as it aimed at replacing cash crops with slower and more moderate returns, and he ran afoul of the farmer's traditional nemesis, the weather. One summer brought such rains that his already depleted fields lost more soil than in any previous season. The lack of rain another year wiped out his winter wheat. While he experimented with other crops and methods in his home county of Albemarle, he continued to raise tobacco in Bedford County, where fully half of his lands were located. But even this could not produce the cash needed to meet his needs, and by the time he left Monticello in 1797 to assume his duties as Vice President, he confessed to his friend James Madison: "Much as I am an enemy of tobacco, I shall endeavor to make some for taxes and clothes."[17] By 1798, his need for cash was such that he resumed the production of tobacco on a full scale.

Jefferson was never able to return to the close personal supervision of his farms until his retirement from the presidency in 1809. By then he was nearly 66, and within a few years, he gladly turned the responsibility for managing his farms over to his grandson. His schemes for reforming his farming practices, while not wholly without result, had largely been frustrated by the need for money. Nor was that the end of the story. His indebtedness continued to increase, and his means for extricating himself diminished. This was because, like the soil itself under constant exposure to tobacco and corn production, the value of his land had steadily eroded in the financial conditions of the early 1800s, so that he found himself laboring under a debt that had been, for the most part reasonably contracted, but which could not be retired without liquidating most of his assets. When he died at the age of 83, his only hope of paying his debts and providing in some measure for his large family was a scheme for a public lottery.

How could one of the most learned and intelligent men of his time, who was wealthy, knowledgeable about farming, and hated debt, come to such a pass? It is, of course, a long and very complicated story, but some aspects of it stand out in relief, particularly in the light of current developments in rural America. For one thing, Jefferson's land didn't all come to him free and clear. The 10,000 acres he and his wife inherited at the death of her father carried a sizeable mortgage. To retire this debt, he sold off about half the land, but he had to accept promissory notes for these lands, which the British mortgage holders would not honor. When the Revolution created an inflation that made Virginia currency virtually worthless, his note-holders paid him off, a state of affairs that left him little better off financially and still owing the British creditors. He thus had to pay this debt twice, which curiously recalls of the plight of nineteenth-century farmers, who were sometimes tricked into doing the same thing by unscrupulous landlords, such as the one in Hamlin Garland's "Under the Lion's Paw."

John C. Miller makes the point in his book on Jefferson and slavery, *The Wolf by the Ears*, that Jefferson was counting on Virginians and other Americans after the Revolution to restrain themselves in their consumption of foreign goods and particularly in availing themselves of credit at the hands of British merchants.[18] His hope was that the austerity and self-reliance of making do in the war years would show Americans that they could get along without the "baubles" of Britain. In this he was bitterly disappointed, and he complained from Paris about this unseemly and disastrous tendency. But not only was he himself, as we have seen, unable to extricate himself from the clutches of British creditors, he was completely unable to live frugally, as he advised. This lack of frugality, as Miller shows, had a good deal to do with his further inability to resolve his slave-holding problems.

He indulged his expensive tastes freely while in Paris (1784-89) and in the years to come, even though he had constantly to borrow money to finance them. Books, he said, were "necessaries of life" for him, as indeed they were. But they were also very costly by any standard, let alone those of rural America, and he imported them in very large numbers. When he sold his magnificent library to Congress in 1815, his serious indebtedness did not deter him from assembling another collection of more than 2,000 volumes. He also imported other costly European baubles such as fine wines, which his family and guests consumed in great quantities. According to his overseer, Edmund Bacon, he was hospitable to a fault, permitting an endless stream of uninvited visitors to impose on the Monticello household and consume its stores. Nor could he deny his many grandchildren the expensive gifts they fancied. Though essentially a man without vices, his virtues were very expensive.

And yet Jefferson seems justified in his belief that his indebtedness, if persistent and considerable, was not in excess of his ability to redeem it in terms of his real property, which was very extensive. The "coup de grace," as he called it, came late in life when he co-signed the notes of a close friend, who promptly went bankrupt and then died. Jefferson could hardly have foreseen, and when it happened could scarcely credit, the devastating drop in Virginia land values that occurred as a consequence of the economic conditions leading up to the panic of 1819. Having to borrow money just to pay the sharply-rising interest forced him into a position where his only financial way out was to sell land, but these same economic conditions and cheap lands in the West made this impossible. "The only relief I wanted," he lamented, "was a market for property, where it might be sold at a fair price and effect the payment of my debts..."[19]

All of this calls forcefully to mind the situation of so many farmers in the Midwest in recent years who, having been advised to "get big or get out," contracted loans that were apparently prudent, based on the rising land values. It was the collective misfortune of these farmers to be caught by calamitous declines in the prices of both crops and land, coupled with record high rates of interest, costing many of them their farms and everything they had worked for. Their dismay at the disappearance of fair prices and their sense of having been tragically victimized are surely as justified and as well-founded as Jefferson's. And like him, they can hardly be blamed if they are less concerned with conserving the soil than with saving their farms.

There are admittedly limits to this parallel, but the similarities are suggestive. Just as Jefferson's lavish lifestyle was a contributing factor in his financial difficulties, so the modern farmer and his family cannot do without expensive goods and services that were unheard-of luxuries to previous generations, not to mention the exorbitant costs of state-of-the-art farm equipment. Where the affluent society sets the standards, a life of frugality, true husbandry, and self-reliance seems no more possible for them than it was for Jefferson. If the conditions of American life have not been propitious for the agrarian ideal he envisioned, they have been rife with the perils of prosperity. As expectations rise and rural life becomes more closely modeled after its urban counterpart, farming becomes more precarious and claims ever more substantial victims. Though he had foreseen that Americans would prove to be more interested in making money than in securing the intangible blessings of rural life, Jefferson apparently did not foresee that even so formidable a farmer as himself was vulnerable to the hazards of agricultural economics. In what is perhaps a parable of our culture, the fate of the American farmer, the special object of Jefferson's hopes for a nation worthy of a New World, has become disarmingly reminiscent of the failures and misfortunes of Jefferson himself.

Notes

1. Thomas Jefferson, *Writings*, Merrill D. Peterson, ed., (New York: The Library of America, 1984), 290.
2. Ibid., 290–91.
3. Ibid., 290.
4. Ibid., 291.
5. Ibid., 836.
6. Ibid., 301.
7. Ibid., 836.
8. Richard Hofstadter, *The Age of Reform: From Bryan to F. D. R.* (New York: Alfred A. Knopf, 1968), 23.
9. Wendell Berry, *The Unsettling of America: Cultural & Agricultural* (San Francisco: Sierra Club Books, 1977), 7–8.
10. George Fredrickson, "Down on the Farm," *New York Review of Books* (23 April 1987): 37.
11. Berry, *Agriculture*, 42.
12. Aldo Leopold, *A Sand County Almanac* (New York: Oxford University Press, 1949), 119.
13. An excellent book about the transformation of American agriculture is John L. Shover's *First Majority—Last Minority* (DeKalb: Northern Illinois University Press, 1976).
14. Jefferson, *Writings*, 842.
15. Ibid., 293.
16. Ibid., 293.
17. Dumas Malone, *Jefferson and the Ordeal of Liberty* (Boston: Little, Brown and Company, 1962), 205.
18. John C. Miller, *The Wolf by the Ears: Jefferson and Slavery* (New York: The Free Press, 1977).
19. Thomas Jefferson, *The Works of Thomas Jefferson*, Paul Leicester Ford, ed., 12 vols. (New York: G. P. Putnam's Sons, 1904), vol. 12:462.

Energetically Thinking About The Environment

Charles Schweighauser

ENERGY IS THE PRIMUM MOBILE. Without energy, life would not function, movement of non-living physical matter would cease, the sun would not shine, and the night sky would be black. Without energy the universe is impossible.

Energy efficiency is defined as the ratio of useful energy or work that comes out of a system to the amount of energy supplied to the system. The system can be a machine for which the energy source is a fossil fuel such as gasoline, kerosene, or coal, or it can be a naturally occurring system such as a planet in orbit or the sun. The system, of course, can also be a living organism.

As we will see when we discuss the laws of thermodynamics, no efficiency ratio can reach one hundred percent. There is never as much work coming out of a system as energy going into the system because some energy must always be converted to heat. As a matter of fact, efficiencies of most systems are very low. Plants use solar energy in the photosynthetic process at efficiencies that vary between one and three percent if all the wavelengths of incident radiation at the surface of the earth are considered. However, it turns out that natural systems use solar energy at considerably higher efficiencies than humanly-contrived technological systems use energy from other sources. It is this lag in technological energy efficiency that is responsible for part of our environmental problems.

BIOLOGICAL ENERGY

In order to arrange our thoughts properly, we will distinguish between two kinds of energy sources: biological energy sources and technological energy sources. The first of these sources is defined as the energy source necessary for maintenance of biological systems. The ultimate energy source used by most living organisms is the sun. The sun also supplies energy for all of the earth's naturally occurring physical systems that influence biological behavior. Weather changes, for example, are the basis for many of the patterns of organic activity. Diurnal and even hourly changes in organisms result from changes in the sun's position and amount of cloudiness. Precipitation is the direct result of solar energy, and in turn helps determine quantities and qualities of plant and animal life in all ecosystems. There are enormously complex interrelationships triggered by solar energy as it powers both physical and biological systems.

Biological energy may be characterized in several ways, including the idea that it is nearly spontaneous. Biological energy passes into plants by the photosynthetic process and is then available to herbivores, usually by only a few grazing manipulations. A plant captures solar energy by photosynthesis, using several elements from the natural world such as nitrogen, phosphorus, hydrogen and oxygen (in the form of water), carbon, and trace elements to make primary plant production, accumulating biological energy in the form of chemical bonds. A herbivore eats a plant, and in turn may be captured and eaten by a carnivore. Rabbits eat the accumulated biological energy of plants, and foxes obtain biological energy stored in rabbits. There are thus few manipulations of tissue by the rabbit, other than grazing for and eating plants, that are required to obtain biological energy.[1]

We may further observe that biological energy poses little stress on living organisms with which they cannot cope because all living matter evolved with solar energy, and thus has adapted to it and is able to exploit it.

TECHNOLOGICAL ENERGY

Technological energy sources are used in activities that are not essential for maintenance of biological systems. Technological energy is characterized as not being spontaneous because its source must be manipulated before it can be used. Coal must be dug out of the ground; oil must be drilled for, piped, and refined before it is useful in automobile engines; and uranium must be mined and concentrated before it can be used in reactors. Technological energy sources almost always require many manipulations before they can be used. One would be hard pressed to think of all the manipulations that are necessary to provide gasoline for automobile engines: acquisition, transportation, handling, refining (including rearrangement of molecules), addition of chemical compounds, storage, transportation to dispensing sites, and final handling to put gasoline into the engine of an automobile require hundreds of manipulations, all of which require energy. Compared to the highly complex, manipulative technology needed to acquire technological energy for an automobile, a horse eating captured biological energy of the plant material it feeds on is a simple, nearly nonmanipulating transportation system. A horse can only travel 30 miles per day, is sometimes ill (but cars have flat tires), and is uncomfortable to ride on when the weather is cold or wet; but a horse is also a much more energy efficient transportation system.

Before the Industrial Revolution began in a large way in the eighteenth century, manipulations required for technological tasks were performed either by natural energy flows, such as windmills and water wheels, or by natural biological energy flows, exemplified by human and animal muscles performing farm work, construction work, and other activities.

The Industrial Revolution added a new dimension to the use of energy for technological purposes in the sense that technological energy sources were employed for the first time to power not only technological devices but also to acquire technological energy sources. For example, coal was mined with the help of steam engines that used coal for a fuel. The Industrial Revolution used technological energy sources such as coal for acquiring more of the same resource. Acquisition of technological energy based on biological energy contained in muscles of animals and human beings was gradually replaced by machines.

The Industrial Revolution made possible the use of large reserves of technological energy sources which in turn allowed continuously accelerating growth of technological energy. As the Energy Revolution (a more accurate term for the Industrial Revolution) progressed, it moved from use of accessible energy sources, such as coal, to more remote energy sources when the easily accessible sources were depleted. Coal was the resource that sustained the early years of the Energy Revolution. About a hundred years ago oil and gas became the fossil fuels on which industry and technology sustained themselves, together with coal. Oil, to be useful, however, generally requires more manipulations than does coal, so that more technological energy must be put into oil-based technological systems than into coal-based technological systems. Today we are faced with having to use even more remote energy sources, thus forcing us to go to Alaska or the Persian Gulf countries for oil. But as known reserves of oil can keep our growing industrialization running for only a few more decades, we find ourselves in a bind about where to obtain technological energy sources for the future. Thus, we turn to atomic energy to meet our demands, at least for electricity.

It turns out that the materials needed for fission, such as uranium 235, are much more inaccessible than any of the fossil fuels in terms of the number of manipulations required to make them useful for producing energy. The more inaccessible the technological energy source, the more manipulations are necessary to make the source accessible, and consequently more technological energy is required to bring inaccessible sources to usable states. Indeed, it is interesting to calculate the net efficiency for production of electricity by atomic fission. When one considers not only the large amounts of technological energy that must be used to bring uranium into concentrated enough form in order that it can be used in fission processes, but also the energy that goes into building and operating reactors, generators, and other machinery in the entire plant, the transmission systems, together with the energy necessary for processing and storage of spent radioactive fuels, and the energy required for cleaning up accidental messes (Three-Mile Island and Chernobyl, for example), the conclusion might be that more technological energy has gone into the system than is obtained from the system. The extra energy that goes into the system to make up the deficiency is probably supplied primarily by fossil fuels and secondarily by hydroelectric power. Thus, it may be that the atomic energy industry is currently based on fossil fuels and water power. When we run out of coal, oil and gas, which we will quite likely do in the not too distant future, it may be that the entire atomic energy industry will in turn be threatened. It should be pointed out that no new atomic power plants have been ordered in the United States since l978; part of the reason may be the energy inefficiencies of atomic power.

We can further characterize technological energy sources by noting that they stress the environment because these energy sources did not participate in the evolutionary history of plants and animals in ways in which these sources are now being used. Manipulation of energy sources that are less spontaneously available to do work compounds our environmental problems: the greater the separation of energy sources between their end use and their natural states the more the energy problem is compounded and the more environmental problems accrue. Technological energy sources that require few manipulations, such as wood, peat and natural gas, are now inadequate to provide easily accessible technological energy for our industrial societies and technological life-styles.

There are other ways in which technological energy sources differ from biological energy sources. For example, the source of biological energy, the sun, is remote from the earth. The sources of almost all of our technological energy are in or under the surface of the earth. In order to use them we often not only have to tear up the earth, thus disrupting in large-scale ways many ecosystems, but we also go through many manipulations to make these sources usable. On the other hand, nuclear fusion manipulations needed to produce solar energy are taking place 93 million miles away in the sun. Solar energy received on the surface of the earth is ready for use by photosynthesizing biological systems without further manipulation. No technological energy source is as spontaneous and available for use as the solar energy that flows into biological systems. This is the significant and ultimately crucial difference between biological and technological energy sources. Put another way, it is the amount of technological energy that we use that disrupts and redirects the flow of biological energy originating in the sun that determines environmental stress.

THERMODYNAMICS: THE LAWS OF ENERGY

Thermodynamic considerations are of fundamental importance to understanding the operation of the natural world. Thermodynamics is the study of the relationship between heat and other forms of energy. Indeed, it was the recognition in the nineteenth century that heat is a form of energy that first gave impetus to the science of thermodynamics.

The first law of thermodynamics states that energy is neither created nor destroyed, and is sometimes referred to as the law of conservation of energy. Energy can exist in many different forms, but it ultimately becomes heat energy, a form that is useless for transformation into metabolic processes in biological systems.

Another way to state the first law of thermodynamics is to say that for changes in an isolated system, the energy within that system always remains constant. An isolated system is thus defined from the first law point of view as one which is energetically self-sustaining. This way of stating the first law says nothing either about the form the energy is in or about the rate of changes in the system. Now let us see if we can think of an isolated system. We have already mentioned the intricate system of gasoline production for an automobile. Thus, we could hardly regard an automobile as an isolated system, cut off from and independent of the rest of the environment. We know that there are no isolated biological systems either, for all living systems depend on the input of biological energy which is ultimately traceable to the sun. As a matter of fact, no matter what system we think of, whether it is biological, physical, or technological, we soon discover that there is no thermodynamically isolated system. There is one exception to this statement: the only thermodynamically isolated system that we know of is, by definition, the universe. Except for the universe, isolated systems cannot occur either spontaneously or by contrivance.

If one wants to play god, one can try to create one's own universe, which is what technological optimists try to do. One will soon discover, however, that no biological, physical, or technological system can operate indefinitely without an input of energy from the rest of the universe. Translated into environmental terms, we had better learn to cherish what we have, for there is no way that we can create our own environment or ecosystem that is energetically as efficient as the natural world. The one we live in and depend on is the best one we will ever have from a thermodynamic point of view. There are no energy efficient substitutes.

The basic law of the natural world is paraphrased in John Muir's observation that "everything is hitched to everything else in the universe." This is not a preservationist's abstract thought but a firm physical condition based on the physics of energy. Energy flow is not just a theoretical or hypothetical construct but a physical phenomenon that can be measured with precision.

The second law of thermodynamics tells us more about constraints in the natural world. All systems, whether they are biological, physical, or technological, are by definition dynamic systems. It follows that any dynamic system must be a thermodynamic system, that is, an energy-transforming system. The second law of thermodynamics tells us that in every dynamic system, some of the energy cannot be converted to useful work but must be degraded by the system to energy in the form of heat. Heat cannot be metabolized by biological systems, and its loss is quite often a nuisance in technological systems.

Heat that escapes from any system can never be recaptured and is lost forever. Furthermore, the energy that does produce useful work is lost from the system once that work has been completed. To keep any system functioning, a constant input of energy is required. Thus, energy flows through a system, never to be recaptured or reused. Useful energy is truly consumed, never recycled to a useful form, although the sum of all energies remains constant in an isolated system according to first law. Any system in which energy is operating, therefore, must be regarded as an open system.

Because energy flows through all systems, regardless of whether they are biological, physical, or technological, it is lost as heat. Electricity flowing in a wire causes the wire to gain heat. Rub your fingers quickly and firmly over the top of your desk: energy is converted to work (motion) and heat, as you will quickly discover. Eventually, all energy in a system is degraded to heat, and thus all useful energy in the universe doing useful work will end up as heat. This situation, agreed on by many thermodynamicists, theoretical physicists, and cosmologists, is the so-called "heat death" of the universe. When high quality electromagnetic radiation from stars has been changed to heat (that is, when stars cease to shine by nuclear fusion processes), when planets, stars, and galaxies have lost their mechanical motion due to friction in interplanetary and interstellar media and their energy of motion is converted to heat, the universe as a whole will be dead.[2] The universe will have reached thermodynamic equilibrium when there are no more sources of energy of any kind.

Thermodynamic equilibrium for any system—physical, technological, or biological—is death because the various kinds of energy they convert to work will have been degraded to heat. No one knows when the heat death of the universe will occur because thermodynamic theory is concerned with direction and not rates. Our own star, the sun, is relatively stable in its present energy mode (i.e., hydrogen to helium fusion that creates radiative energy) for several billion more years.

The second law of thermodynamics can be stated in a number of ways. A formal way of stating it is to say that it is impossible to operate any system in which the net effect is absorption of energy in that system at constant temperature which results in the production of an equivalent amount of work, i.e., an efficiency of one. A more informal expression of the second law of thermodynamics is: "There isn't any free lunch," meaning that if one is going to use energy to do work, the natural world demands payment for use of its energy by taking it back in the form of heat which in turn means that high quality energy is no longer available to do work. It is interesting to think of what would happen if the second law of thermodynamics were not in effect. Biological systems, for example, would not age because energy would recycle; we could live forever. On the other hand, there would be no human beings because if energy recycled there would be no evolution. Without evolution, life would not have started in the first place, and, carrying this idea to its logical conclusion, the universe itself would not be possible. Rivers would flow uphill, automobiles could be operated forever on a constantly full tank of gasoline, perpetual motion machines would be the order of the day, and we might be able to recreate dinosaurs for our hunting pleasure. As the universe really works, however, we constantly have to keep filling our fuel tanks if we want to drive our automobiles, rivers always flow downhill, and we will never see live dinosaurs.

One can speculate on the merits of the natural world as it is and as we might like it to be, but a far more useful exercise is to try to understand it as it does exist and to learn how to live with that reality in lieu of an appealing but unworkable technological construct. Too often we are presented a vision of what the world will be like when we have an advanced technology that will solve all our problems.

But if we try to live with this kind of a situation we will sooner or later have problems with the natural world, a situation that is now emerging. We can thus define environmental destruction as the interruption and diminishment of biological energy by the over-zealous use of technological energy.

Energy that operates in biological, physical, and technological systems is called free energy because it is available to dynamic systems. After free energy has passed through a system and is no longer available to do useful work, we say that the entropy of the system, which is a measure of disorder and randomness, has inevitably increased together with the inevitable increase in heat. We can regard entropy as a measure of energy that is no longer available to a dynamic system, and since all systems represent order, entropy must be a measure of disorder and randomness.

For any irreversible process, which is the way the natural world works no matter what kind of system is involved, the sum of all changes in entropy is positive. That is, entropy in an isolated system always increases, which is perhaps the most direct and succinct statement of the second law of thermodynamics that we can make. Entropy and the second law of thermodynamics tell us that there are no reversible processes in the natural world. A corollary statement to the second law is that entropy of a system does not spontaneously and permanently decrease. This statement simply means that unavailable energy does not become available again. Thus, there are no changes in the natural world in which the entropy sign is permanently negative.

When a plant photosynthesizes solar (biological) energy and uses this energy to grow and maintain itself, there is a temporary increase in order, organization, and structure of the organism and a temporary decrease in entropy and disorder. At first, this statement may seem like a contradiction of the second law of thermodynamics, which says that entropy always increases. We must remember, however, that we have only a temporary decrease in entropy, and, more importantly, we have neglected the rest of the environment. Energy that makes possible a temporary increase in order in the plant is free energy that is being degraded into entropy not only as the plant respires but also into heat at the source of the free energy, which is the sun. Entropy of the total system, made up of the sun, the plant, and the surroundings (earth, interplanetary space, and in a real sense the rest of the universe) increases and thus provides a temporary decrease of entropy and disorder in the plant. Put another way, the surface of the earth is highly organized by biological systems, but at the expense of the increase of entropy of the universe.

Thus, entropy of any system can only apparently decrease if we disregard the environment of the system. If we regard any system as thermodynamically isolated from the rest of the universe it will appear as if that system can be maintained and ordered forever. As we have seen, however, there are no thermodynamically isolated systems—biological, physical, or technological—in the universe. Everything is hitched to everything else.

Unfortunately, we tend to think of our technology as being energetically and thermodynamically isolated from the rest of the environment, giving us an illusory "technological fix." For example, many people have the notion that technology will rid water and air of our pollution. Technology can certainly clean up a river, but only at the expense of energy and materials used to design, build and maintain technological systems that have come from elsewhere in the environment and therefore stress some other part of the environment. Entropy can be decreased temporarily only in a local area at the expense of a permanent increase of entropy in the total environment. There are no unilateral actions: we can never do only one thing at a time because we do not live in a thermodynamically isolated system. If we are to clean a river of its pollutants, increasing entropy by technological gadgetry is no solution. A river can only be cleansed by a halt or at least a significant cutback of the pollutants going into it in the first place and by letting natural energy flows, natural flushing as the result of precipitation, run through the watershed.

There is another way to think about technology and the environment. The more sophisticated a technology is, the more technological energy it uses. Further, sophisticated technologies require more simplified (and hence ecologically unsound) systems in which to operate. For example, an ax is a simple, unsophisticated technological device that requires little technological energy to make and none to operate. It functions quite well in a highly diverse ecological system such as a forest that is dominated by biological energy flow from the sun.

A lawnmower, however, is technologically more sophisticated, uses more technological energy to design, build and operate, and is not useful in a forest. It requires a simplified, early successional primitive ecosystem—a lawn (and hence diminished biological energy flow)—to be operative.

An automobile is more technologically energy demanding than a lawnmower, and is unthinkable in a highly complex biological energy flow system such as a forest. It requires an even more simplified environment than a lawnmower in which to operate. This environment is the interstate highway system, a setting suitable for automobiles first and for biological organisms, including human beings, secondly.

Airplanes are quite complex technologically, require enormous inputs of technological energy, and thus can operate only in highly simplified environments. O'Hare Airport in Chicago is designed for the operation of airplanes and automobiles, and only incidentally for biological systems called human beings.

Put another way, the more energy intensive a technology is, the more technological energy it requires, the farther it is removed from and the more it interrupts biological energy flows, and thus the more simplified its environment must be—and the more impoverished the biological world becomes.

We are thus led to an important generalization about our participation in the natural world. We can say with assurance based on the second law of thermodynamics that there are no ecologically sound technological solutions to our environmental problems. We cannot build a machine or a black box that will reverse pollution without creating its own pollution. It is impossible technologically to restore a river to its precisely pristine state without producing changes elsewhere.

The second law thus suggests that our pollution control laws—The Clean Air Act and the Clean Water Act, for example—will not work.

There are solutions, some of which have a technological basis, to environmental problems that are more ecologically sound than other technological solutions because they use less technological energy more efficiently. It is the ecologically sound solutions that we must learn to recognize and to employ as we seek wise participation in the natural world. One of the more sound (because it requires a small amount of technological energy to fabricate, distribute, and assimilate) technological solutions to environmental problems is birth control; such techniques, if properly used, provide one of the technologies necessary to diminish growth of the human population. If there are fewer human beings, there will be less technological stress on the natural world providing, of course, that the per capita consumption of energy does not increase as the human population declines. This kind of approach is more environmentally acceptable than the expanded and therefore entropy-producing technologies that are proposed to house, feed, and clothe an exploding population.

There are other implications of the second law of thermodynamics for the natural world. High entropy and disorder represent maximum probability, the final state of any system. Conversely, a condition of low entropy and low order represents a condition of minimum probability. Therefore, the more technological energy we use, the faster we are pushing our environment to a state of maximum probability and hence maximum disorder. We are, in short, accelerating the increase of entropy. Ultimately, all pollution can therefore be regarded as entropy pollution.

We can summarize the second law of thermodynamics and its implications for the environment by suggesting that it can help us to define the practical limits of energy efficiency of both biological energy and technological energy. We can use the second law to help us know how better to live with food resources, for example, and how to use less technological energy in order not to stress the environment irreparably.

If a system has high entropy, we have seen that the system will also have a great deal of disorder, and we will not be able to discover very much about its details. Conversely, if a system has low entropy, there will be more order and organization and we should be able to find out a great deal of information about it. However, even in ecosystems that approach climax conditions, where solar energy is used primarily for maintenance and not for biological production, and hence organization is greater and entropy is low, the amount of complexity is so great that all we will ever have is a shadow of precise knowledge about the system.

Furthermore, knowledge about any system can never permanently increase since all systems move toward maximum entropy and hence disorder. Only reversible processes permanently gain order, and, as we have seen, based on the second law of thermodynamics, there are no reversible processes: time and energy move in only one direction.

One might argue, however, that one's own understanding of the natural world has increased, for example, by reading this chapter. True enough, but this increase is only a temporary increase: one certainly will not live to perpetuate and to add to one's knowledge forever. Furthermore, the reader is alive and metabolizing, which in turn means that he or she is using biological energy from breakfast, lunch, and dinner to stay alive and alert in order to read this book, but only at the expense of the increase of entropy elsewhere in the environment. Temporary increase of knowledge and order is possible due to the permanent increase of entropy and disorder in the total environment. Temporary increase of order in one's brain is inexorably accompanied by a permanent increase of disorder throughout the rest of the environment: the reader and his or her brain are not thermodynamically isolated systems.

One might also argue that there can be a permanent increase of knowledge in another way, and that is by putting this book in a library we can preserve its knowledge forever. However, books in libraries decay due to natural processes of energy flow. If books decay in the library rapidly due to air pollution, we are only compressing nature's time scale because we are using technological energy sources to manufacture these pollutants. Accelerating the increase of entropy creates a condition of rapidly increasing disorder.

There is one other way to distinguish between biological and technological energy. Biological energy operates in such a way that probabilities are ordered for the survival of an organism and of a species. Technological energy, on the other hand, operates in a very narrow range of probability in order that putative certainties and absolutes determine the behavior and activities of the system. When an environmental condition imposes itself on a technological system for which that system has no information, the system breaks down. For example, an automobile cannot function under water. In the natural world, however, only rarely will environmental conditions be such that a biological system cannot find somewhere in its information bank a combination of probable information that will allow it to survive and to adapt to or take advantage of changed environmental conditions.[3]

Put another way, technological energy systems are predictable only to a point, and therefore unreliable (e.g., when an automobile breaks down). Biological energy systems are also only partly predictable, but almost always reliable. By substituting technological energy for biological energy, we are trading a sense of comfort and security in the biological world for a false security, the unreliability and frustrations of the technological world—frustrating when we find the latter is neither predictable nor reliable.

The amount of information in a humanly-contrived, technological system is low compared to a biological system. The amount of information one has about a humanly-contrived system, compared to the amount of information available, is quite high. We can know nearly everything there is to know about an automobile or building because we selected the information for them when we designed and built these technological systems. Thus, the nearly one-to-one ratio between information available and information perceived in technological systems gives us the illusion of one-hundred percent reliability and predictability.

Biological systems, however, are quite reliable because they contain a diversity of information. Reliability increases compared to technological systems.

Life can also be defined in terms of thermodynamic considerations. Organisms must accumulate information in their gene structures in order to exist. They must organize themselves around information that has been proven through evolutionary trial and error to be useful for survival. In order to accumulate information, which is a temporary reversal of entropy and therefore a precarious existence in a state of least probability, organisms must consume free energy. Therefore, life can be defined as a temporary accumulation of information and free energy in a temporary and improbable state. Life is possible only because there is an increase of disorder and probability in the total system. The universe not only supplies energy for biological systems via the sun but also is the entropy sink needed for life to exist. Since the universe is tending toward the most probable state of disorder and maximum entropy, and life is the temporary denial of inexorable thermodynamic decay, all life will eventually come to an end.

Erwin Schroedinger in his wonderful little book, *What is Life?*, uses the term "negative entropy" for free energy since positive entropy is what increases after the expenditure of energy. He suggests that it is "negentropy" that all living organisms feed on to stay alive. Since all systems tend toward thermodynamic equilibrium, Schroedinger notes that life may be defined as resistance to thermodynamic equilibrium. Life thus is avoidance of one's own entropy for as long as possible; as noted previously, death occurs when an organism—or any system—reaches thermodynamic equilibrium.

Increase in entropy is a necessary result of life. Life and death are inextricably bound together by energy flow: negentropy—free energy and the temporary accumulation of information—is life, but life degrades energy to positive entropy, disorder, and death. Life and death are tied together in a relentless thermodynamic relationship.

An indisputable corollary to this idea is that if we human beings persist in accelerating the increase of the entropy of our environment through the excessive use of technological energy sources, we are accelerating our own demise, directly by altering our environment at such a rate and to such an extent that we may no longer be able to survive in it[4] and indirectly by causing the extinction (maximum entropy) of the rich diversity of plants and animals on our planet. It would therefore seem that we should design our technology to use as little energy as possible in order to harmonize better with natural biological energy sources and flows.

Notes

1. It may be argued that mushrooms and other non- photosynthesizing organisms are not alive according to our definition because they are not photosynthesizing plants. Ultimately, however, energy that nourishes them is the result of solar radiation acting on materials that are in turn used by fungi and other non-photosynthesizing organisms.
2. The postulated heat death of the universe is an extrapolation based on knowledge of small-scale events. The heat death of the universe seems plausible but, of course, cannot be taken as an absolute.
3. One of these rare events would be the rapid demise of the dinosaurs and other organisms at the Cretaceous-Tertiary boundary.
4. What I previously termed entropy pollution government agencies call air and water pollution, toxic and hazardous substances, and solid and radioactive wastes.

A Secret Power: Symbolic Bears in Contemporary Culture

Elizabeth Atwood Lawrence

(Reprinted with permission of Bowling Green State University Press and the author.)

VIRTUALLY EVERY ANIMAL with which people interact has, in addition to its own physical being, a symbolic identity which may or may not accurately reflect the animal's nature, but which arises out of individual and cultural perceptions of that species. Study of the symbolic aspects of animals can shed light on the deeper dimensions of human-animal relationships and help to explain why certain species have occupied such an important role in human life and culture. Bears, for example, have been a source of fascination in many societies for countless ages. As large, spectacular creatures, bears have often commanded awe, respect, and admiration, as well as evoking fear and antagonism.

Since the dawn of human consciousness, wherever the bear has been encountered he has been the object of special attention. As Paul Shepard has observed, "In his liveliness we sense a secret power, as though he were the embodiment of our anima—spirit, soul, or holy breath."[1] Bears may have inspired humankind's first mystical feelings and concepts, for according to Joseph Campbell "the earliest evidence anywhere on earth of veneration of a divine being is in the Alpine bear-skull sanctuaries of Neanderthal man." Through studying such areas, archaeologists have found indications that during paleolithic times the remains of the now extinct cave bear were used in worshipful rites and ceremonies.[2] Ancient peoples identified huge bear images in the night sky and named two great constellations, Ursa Major and Ursa Minor, in honor of the awesome creatures. These configurations—the big and little bear—are two of the oldest star groupings known, and were recognized long ago by such divergent societies as the North American Indians, the Basques, and the early Chaldeans.[3]

In numerous non-industrial societies of the world, extending into the present, attitudes of esteem and reverence for bears are recorded as important cultural elements. Bear ceremonialism was once widespread among certain peoples of North America, Asia, and Europe[4], and has persisted into modern times.[5] Bears were admired for strength, courage, and the special wisdom that allows them to live all winter in hibernation without eating.[6] Though bears were hunted and eaten, special rites were performed to ensure respect and proper treatment for the slain ani-

Originally published in *Dominant Symbols in Contemporary Culture.*

mals. Contemporary Koyukon Indians of Alaska still retain a cultural tradition of deep respect for bears, regarding them "near the apex of power among the spirits of the natural world."[7]

Belief in and admiration for the supernatural power of bears and desire to obtain these powers resulted in formation of bear cults among Plains Indian tribes in the 19th century. The Ute Indians performed an annual sacred bear dance, paying tribute to "the bear as the wisest of all animals" who possessed magic and medicinal power. Dancers were able to send messages to the dead and some times could obtain curative abilities through the agency of bears.[8] The animals have often been associated with healing. The Cheyenne held that a bear can cure itself and other bears.[9] The Lakota "believed that their knowledge of healing wounds had been taught by the Grizzly bear," and in a healing ceremony the medicine man and his helpers dressed and acted like bears in order to cure the sick.[10] Among certain Gypsies as well, bears have magic power for curing and alleviating suffering.[11] Even today, especially in Asia, certain bear products such as bile from the gall bladder are believed to have medicinal value.[12]

In some cultures, actual kinship between bears and people is recognized. The Utes, for example, acknowledge bears as ancestors. Northwest Coast mythology includes intermarriage between human beings, with offspring having the characteristics of both species. There is a belief among Gypsies that the bear came into existence through birth to a virgin girl and thus has a close association with humankind.[13] Virtually all people who live in contact with bears recognize the animals' similarities with human beings. Bears walk upright, appear tailless, are omnivorous, exhibit facial and bodily expressions resembling people's, may whine and shed tears in distress, are devoted mothers to their playful and dependent young, and seem to behave with conscious intelligence. Even objective scientists attribute conceptual abilities to bears, and grizzlies have been described as the only animals to understand that their tracks reveal their movements and take pains to conceal them.[14]

The Anglo-Europeans who came to explore, conquer, and settle in the New World wilderness, however, brought with them an entirely different ethos; bears were generally seen as antagonists. They were formidable predators, an intrinsic part of the obstacles that had to be overcome in the untamed new land. Known for their courage and ferocity, bears came to represent the challenge of a harsh life that brought the pioneers into daily contact with the wild elements that they sought to tame. Not just a matter of survival, killing a bear became a ritual of manhood, a way for frontiersmen to prove their strength and virility—and to assert human supremacy over nature. Bears were perceived as enemies who threatened the progress of civilization, the working out of Manifest Destiny on the American continent. As animals not amenable to human domination, bears' perceived unpredictability and uncontrollability often evoked hostility and made the animals appear symbolic of all in nature that is yet untamed.[15]

It is not coincidental that a change in attitude toward bears began to be apparent around the time of the turn of the century. In 1890, the American frontier of settlement had been declared officially closed. Shortly thereafter, historian Frederick Jackson Turner put forth his now-famous thesis, *The Significance of the Frontier in American History.* Turner asserted that the frontier had been the most important factor in shaping American character and history. From the wild and rugged land itself, he argued, had come the main stimulus for development of the strength, self-reliance, and individualism that characterize Americans and formed the basis for the growth of American democracy.[16] Once the American empire extended to the Pacific and the domination of nature on the continent was established, the wilderness gradually came to be viewed as a value to be cherished rather than merely a resource to be exploited. Though attitudes toward wild animals were still ambivalent, they began to be viewed somewhat more benignly, and even the ferocious bear could be seen, at least by some Americans, with an appreciative eye.

Probably no person represents that newly emerging dual view of the wild in which the urge to conquer was counterbalanced by a simultaneous admiration than Theodore Roosevelt. Though the twenty-sixth President of the United States (1901-1909) was an enthusiastic big game hunter, whose countless trophies attested to his prowess in killing animals, he was at the same time a dedicated conservationist who was instrumental in preserving American wilderness and wildlife. Interestingly, Roosevelt expressed agreement with Turner's thesis about the significance of the American frontier; he wrote to the historian "I think you have struck some fist class ideas." Roosevelt had in fact anticipated some of Turner's concepts in his own book, *The Winning of the West*, published in 1889. As a champion of the strenuous outdoor life and pioneer virtues, Roosevelt was alarmed by the realization that the frontier had vanished, fearing the detrimental effect of this loss upon the vigor, virility, and greatness of the nation.[17]

An incident that occurred during one of his hunting exploits forever links Theodore Roosevelt with American bears and created one of the most powerful and enduring symbols in American culture. In 1902, the president went on a bear hunt near Smedes, Mississippi. Sources differ as to the official reason for his presence in that area, citing it as either "a touchy political situation involving racial equality" or a boundary dispute between Mississippi and Louisiana. After several days of unsuccessful hunting, a bedraggled black bear was run down by the dogs, hit on the head with a rifle, and then roped and tied to a tree. The president was called upon to kill the bear, but he refused to shoot it under such unsportsmanlike conditions. He reportedly told the guide to put the animal out of its misery, and the bear was dispatched with a knife. The next day the story of the president's refusal to shoot the captive quarry was reported in the newspapers, and several cartoons appeared depicting the bear fist as a full-grown animal and soon afterward as a small cub. Clifford Berryman of the Washington Post immortalized the incident in consecutively published cartoons entitled "Drawing the Line in Mississippi"—the caption referring both to Roosevelt's distinction between good and bad sportsmanship and his involvement with the "color line" of the racial issue or the state boundary line.[18] Most significantly, when examined carefully the bears of the two cartoons are quite different. In the earlier depiction, the realistic-looking adult bear is large in proportion to its captor, and defiantly tugs against the rope tied around its neck. In the subsequent version, the bear has been transformed into an appealing little cub, its eyes wide with fright. The young animal, unresisting, is pulled much closer to the man who holds the rope.[19]

Drawing inspiration from Berryman's cartoons, Morris Michtom, operator of a Brooklyn toy and candy store, created a stuffed bear for the Christmas season. Allegedly, he requested and received permission from the president to call his product "Teddy's Bear," a name which soon became teddy bear. Sales were so good during the first year that Michtom formed a new larger company, the Ideal Toy Company.[20] In 1903, the Steiff family, German toy manufacturers, also produced stuffed bears which were fist marketed in New York, and which contributed to the widespread popularity of teddy bears. The bear soon became Roosevelt's mascot and a symbol of his presidency. A series of cartoons featuring the adventures of two bears known as The Roosevelt Bears appeared in twenty newspapers, including the New York Times, and was later published in book form. Almost all cartoons depicting the president included the bear, generally a small cub, as a character. During his presidential campaign, teddy bears were featured in special pins, buttons, and posters advocating his election.

Soon after their creation, teddy bears gained immense popularity that never diminished, and they have always occupied a very special niche in the lives of countless numbers of people. In our society, these symbolic bears often represent an infant's first introduction to nonhuman creatures and serve as many children's earliest and closest companions. Although teddy bears are exceedingly popular with youngsters, they also possess considerable appeal for adolescents, adults, and the elderly. Recently, teddy bears have been successfully employed in many age groups for stress reduction and for various therapeutic purposes. During the last decade or so there has been a remarkable surge of interest in and appreciation for teddy bears. Collecting and restoring old teddy bears has become an absorbing adult hobby, and prices of antique bears have soared. A California enthusiast paid $86,350 for a 1920 Steiff bear in 1989.[21] Numerous books and periodicals on the subject of teddy bears have appeared. Stores devoted entirely to teddy bears have been established throughout the country. Teddy bear sales, contests, rallies, jamborees, clinics, and exhibitions are often held, and clubs for devotees (known as arctophiles) are active. Teddy bear tea parties can be elegant fund-raisers.[22] Currently, the concept of the teddy bear as a companion animal is a truly a remarkable American social phenomenon. Particularly in childhood, but also during other life stages, the toy carries extremely potent symbolic value. Teddy bears are increasingly recognized for their extraordinary functional powers to bring comfort and consolation to people, and are utilized to counteract anxieties ranging from mild routine daily stress to severe trauma and crisis.

Recently I gave a children's sermon in church on the theme of teddy bears. My message concerning the history and significance of teddy bears was received with great enthusiasm. Each child was given a small teddy bear before proceeding to

Sunday School. And on the way out of church adults were offered teddy bears as well, though it was not expected that many would accept them. As it turned out, not one person refused a bear. Men and women alike took special pains, some even aggressively, to claim their bears, and several asked to have extras for relatives and friends. Not one seemed embarrassed or thought the idea foolish; not one believed teddy bears were just for children. All parishioners exhibited broad smiles upon receiving their bears, and immediately held them tightly to their bodies. One elderly woman expressed special joy, and said this was her first teddy bear. The minister revealed that "ordinarily people scatter after church; only ten percent shake hands with me on the way out. Today, every person went through the line to get a bear." A man who usually goes out the side door to avoid shaking hands even jumped the line to ensure getting a bear. One woman who had never before greeted the minister after the service told him, "That's the best service I've ever attended!" She hugged her bear and hugged and kissed the minister as well. "The woman acted giddy, like a child, as though she'd had a drink or two," the minister said. After the service, the janitor asked for and received a bear, then set it down while he began tidying the church. A few minutes later, visibly upset, he frantically sought the minister to request another. "They stole my teddy bear!" he exclaimed. In the following weeks, many comments from church attendees attested to the value placed upon the bears. I learned that one person took the animal to bed each night, another gave it to a sister in a nursing home, who adored it, and several carried it everywhere in pocket or purse, or displayed it on the dashboard of the car. The administrator of a large hospital kept his bear on his desk next to his wife's photograph. "It's there when I need it," he confided. "I need hugs too. I only wish I'd had several teddy bears to help me during the difficult times of the merger a few years ago."

The tension-easing effect of teddy bears was recently utilized by a Rhode Island bus company to combat "busphobia," the irrational fear of riding buses. This fear, felt to be the reason that more people did not use buses, was said to be composed of three main elements: anxiety that "the bus would not go where the rider wants; that the rider will have to sit next to an undesirable character; and that the rider might be embarrassed to be seen riding the bus."[23] A bus company official told me that many citizens had negative perceptions of people who ride buses, believing that passengers are "automatically of a low class, poor, and without much education. The teddy bear campaign was undertaken to stop busphobia and make people proud of their transit system. For six months we gave teddy bears free to anyone who rode the bus. The program was successful; the ridership increased by one percent. The teddy bear is cute, cuddly, and not offensive; it's a positive thing in people's lives."

The same beneficial effect has helped in overcoming "dental phobia," a phenomenon that, according to the findings of the Dental Fears Research Clinic at the University of Washington, is present to some degree in two-thirds of the population. "Bring a teddy bear" is one suggestion made to those who find it difficult to "face the chair" at the dentist's office.[24] Teddy bears are also utilized by hospitals to "introduce children to what goes on in an emergency room in hopes that if they have to come during a real emergency, they won't be as frightened." At the annual "Teddy Bear Clinic Day" at Rhode Island Hospital, for example, youngsters bring teddy bears in for checkups or surgery. On display are x-rays of teddy bears, bears with oxygen masks, I.V. tubes, and casts on their legs, and bears in wheel chairs. The bears' "parents" are allowed to give shots to the patients. Staff members sew up injured teddy bears, replace parts, and prescribe "lots of hugs." As one "grandparent" explained, her child had "been hesitant to go to the doctor, so we thought this was a good idea to come here in a non-threatening situation."[25] At some pediatric hospitals, a teddy bear with the same ailment as the child is used to reduce fear—for example, just before treatment is begun a bear with an eye patch is given to a youngster with an affected eye.

Many charitable organizations throughout the country raise money to provide teddy bears to pediatric wards and children's hospitals for distribution to sick children to aid in easing the young patients' pain and loneliness. Dr. Paul Horton told me of a National Cancer Institute Program in which forty-five countries were involved in providing bears for children dying of cancer. In New York State, the teddy bear campaign is held to commemorate Theodore Roosevelt's birthday, and is sponsored by the National Park Service's Manhattan sites (including Roosevelt's birthplace) and the Theodore Roosevelt Association. During the second annual Theodore Roosevelt Drive in 1987, 10,000 bears were obtained for presentation to hospitalized children at Christmas and Chanuka.

Teddy bears can fill the physical, as well as the emotional needs of some adult patients. At one Florida hospital, all convalescents from open-heart surgery are given a unique bear called "Sir Koff-a-lot." Surgeons explain that following the operation it is very important for patients to cough up phlegm from their lungs, but that doing so causes a lot of discomfort due to the newly sutured wounds. When patients hug the large specially contoured stuffed bears to their chests while coughing, however, the pain is eased. The teddy bears also have been found to "divert the patient's attention from the trauma of surgery to the excitement of recovery" . . . and to provide patients with "the warmth and tenderness they so badly need after heart surgery." The bears have made "a real difference in attitude during recovery," being "not just physically therapeutic but a mental tonic." A nurse noted that "everybody loves teddy bears. Even the most macho male patients come around to hugging their bears in the end. It definitely eases their post-op pain." Patients are extremely enthusiastic about the curative effects of their bears, who wear tee-shirts with the message: "A cough for your lungs /A hug for your heart; /I'll help you bear it/ By doing my part."[26]

Teddy bears have recently found another area in which to do their part—as "recruits" in the police force of many American cities. Officers who are using teddy bears in their daily work report "a teddy bear is quite a tool; it's almost as important as my pin and gun. It has worked miracles." The idea of policemen armed with bears as well as bearing arms began several years ago, and its astounding success has caused teddy bear programs to spread to many areas of the United

States. Requests for help in starting similar projects have come from Australia, England, Scotland, and Germany. The procedure involves police officers giving bears to people undergoing stress and trauma. The teddy bears are generally supplied to the police through fund drives carried out with the help of the local media in which individuals or large companies donate bears or money to buy them. The bears are permanent gifts, and are most frequently given to children, but they are also used for adults.

Teddy bears have proven to be beneficial in any traumatic situation, especially for traffic accident victims, runaways, confused, frightened, and lost people, abused and sexually assaulted individuals, for youngsters who witness a killing or whose parents are arrested, and for children who must testify in court. Policemen say that whenever people are scared, hugging a teddy bear soothes and calms them so that they can communicate with those who have come to help. As one officer expressed it, "In my twenty-seven-year career, I've never seen anything work better. The teddy bear is our emissary, calming the situation down. People may be frightened of policemen, but the bear builds a bridge between the officer and the person, so it helps us to help people." Officers find that "giving a bear to a stressed individual to hold, love, name, and keep works better than anything else. The bear breaks down the language barrier by speaking a common language that says 'here's a friend.'" Police call the bear a "security blanket," and an "ice breaker." They say "it helps make a child understand he is not alone, that the officer cares." The director of one city's "Best Buddy" program told me "at first, no macho cop wanted to carry a teddy bear, but once they discovered what the bear does, they won't go anywhere without it. The bear program has caught on like wildfire."

One police official revealed that "even hard-core street-wise kids with colored hair are intrigued that police have teddy bears. They come up to the patrol cars and look in to see if the bear is really there. They use any excuse to interact with the bear. It's amazing that hardened criminals picked up in a police car say they want to sit in the back seat with the bear. They claim 'I won't hurt it!' The bears create good feelings. One motorcycle cop carries a bear on his bike. Though he doesn't distribute bears, when he gives out a ticket, he still gets smile, even from the person getting the ticket."

Officers describe countless cases that demonstrate the beneficial effects of teddy bears. Following an auto accident, a mother was badly injured and unconscious. Her child, only mildly injured, clung to her hysterically and was totally incoherent. Police were able to "replace mom with a bear," so that the panic-stricken child grabbed the bear, quieted down, and allowed the paramedics to aid the mother. An adult accident victim who was pinned in the car was given a bear to comfort her while a crew worked to extricate her. She took the bear into surgery and still has it in her bedroom. She calls frequently to thank the policemen. Confronted with a badly injured mentally retarded child who would not communicate with the ambulance attendant, police worked through the teddy bear. The child told the bear where he was hurt and thus this information was relayed to the paramedics. As an officer explained, "It's easier to tell your problems to a friendly, fuzzy face." In this case, as in others, officials showed the child what they were going to do by first doing it to the bear. The teddy bear given to one frightened child while in the ambulance became an advocate when police told him, "The bear will stay with you and help you, and will make sure the doctor takes good care of you." Teddy bears are allowed to accompany children when they need x-rays and tests, and they still hold on to the bear when they go home.

Abused children, unresponsive to human contact, react positively to teddy bears. One policeman tells of being called to investigate the case of a four-year-old child who had been severely beaten. "He was terrified and refused to speak. When given a teddy bear, his face lit up like a Christmas tree. It broke the barrier and the child communicated. As a result, we were able to put together a good case for court." Many months later, at the trial, the officer observed the child still clutching the teddy bear he had been given at the time of crisis. "The poor bear was misshapen, where the kid had taken out his aggressions on it by squeezing and punching it, but he still had it," the officer told me. Policemen often find that rape and incest victims, both children and adults, will talk to the bear but not to a person. An investigator revealed that "one twenty-year-old was completely incoherent after a sexual attack, but was calmed down enough by the bear to enable us to get crucial information quickly to help us get the perpetrator."

As a police officer explained the use of bears as fear-fighters for assault victims, "This is where the positive meaning of the program starts to work. You watch tears and sobbing turn into little glistening sparkles in their eyes. They grab the bear and hold on for dear life. The uniformed officer is no longer merely the policeman who is going to help. He just took away some of the pain and hurt, and probably made a friend for life." He finds that "many questions that are sensitive to the victim can be addressed through the bear. The bear acts as the extension of the child. The child starts talking to the bear and in most cases will answer the questions." Bears also provide a source of comfort for a child who has to be taken away from his parents and sent to a foster home, and they give strength to youngsters who must testify in court. "Even when lost kids can't speak English," officers say, "we usually make friends with them through the bear while we make arrangements to help."

Upon entering a home to serve a narcotics warrant, an officer found the suspect's toddler sitting precariously on an outside third-story window ledge. The child was coaxed inside by the officer's promise of a teddy bear. Quick thinking and use of the "Best Buddy" averted a possible tragedy. At the other end of the age scale, an unidentified elderly woman wandered from a nursing home and would not communicate with police who found her. "We were getting nowhere," the officer said, "but when we gave her a teddy bear, it was like turning a key. We got the information necessary to get her back into her facility." A police official also told me about the janitor in his barracks who lost his wife of fifty years. "The man was distraught and unable to sleep. We gave him one of the patrol bears to take home. Next day he came to work more cheerful and said that last night he had been able to sleep for the first time since his wife took ill and passed away. He was greatly comforted by reaching out to touch the bear. It was there for him, friendly and warm. This is an example of our exciting and wonderful bear program."

Participants in the program are always enthusiastic about its success. "Teddy bears are an old tradition," one policeman explained. "Everyone had a bear sometime in their life as a friend to hug, love, cry and laugh with. They're soft and cuddly, with arms long enough to hug back. They're not like a plastic doll. They're warm and loving, not harmful or threatening." An officer with a long career in the police force observed, "Everybody's perception of teddy bears is good. They're important to me. As a child I saw one in a store, begged my parents for it and they bought it for me. It was my companion in childhood, adolescence, and on up. I still have it packed in a case in the attic. It's a sign of love, affection, and warmth. I told that bear every one of my secrets. It's a good thing the teddy bear can't talk." An officer who heads the "Best Buddy" project in a large city found that in addition to making trauma for the victim "more bearable," the program "removes some of the stress a police officer faces every day. Men are proud of their teddy bears because they made someone happy and they could stop some of the hurt. There have even been reports of the police officers giving the teddy bear a squeeze when under stress." A policeman who is dubbed "Officer Teddy" because he often uses the bears says, "Kids love teddy bears. I always did. I still do. I really believe in them." An official of one city's crime prevention program notes that "Americans are slow to touch each other and police hope bears will bring people together more. We are less outwardly affectionate than Europeans. You often see them walking arm-in-arm, but you don't see it here."

City police point out that crime against children is on the rise in this country, so the program is more important than ever. The teddy bear's good image enhances the usefulness of the police force, for it is recognized that police are often at a disadvantage in giving comfort because they are viewed as authoritarian figures. A uniformed officer with a gun is a source of fear, but bears are able to break down the barrier between child and officer. In one city, a policeman who gives out a bear also provides 2 junior police badges—one for the child and one for the bear. The badges reinforce the positive association between the youngster and the police officer as a caring friend. By destroying negative stereotypes, bears are credited with "helping to instill confidence in law enforcement in a whole new generation."

The comforting effect of teddy bears in the lives of senior citizens is also well-recognized. Some charitable agencies have distributed them to nursing home patients and found that they are received joyfully. At the final stages of life, too, bears sometimes fill an important role. Several funeral directors have told me of their experiences with teddy bears. One funeral home keeps a teddy bear handy for visitors. The director explained, "We bought it for children to play with, but we found that adults like it too. They often pick it up and handle it and then hand it on to someone else." Another funeral director revealed that "sometimes as much as twenty years ahead we've had people make arrangements with us to be buried with their teddy bear in the casket." I was also informed by a funeral home operator that "some children are buried with teddy bears, but more frequently the elderly are involved. These people have a fixation on the teddy bear and feel strongly about it. They tell someone else of their request or notify us directly. It's always an old teddy bear that has been in their possession a long time. These people identify with the bear and want it with them in death. It's their friend."

To understand the remarkably potent role of the teddy bear as a lifelong friend and comforter, one must begin with the stuffed bear's origin. There is now definite evidence from a newly discovered photograph that the bear which Theodore Roosevelt refused to shoot was in fact full grown, according to the executive director of the Theodore Roosevelt Association.[27] Why then, did the cartoonist immediately transform the bear to a cub? This action can be explained by the inherent power of juvenile forms (both animal and human) to elicit a positive emotional response from people. The phenomenon is well known and was identified and analyzed by ethologist Konrad Lorenz. Lorenz's studies showed that the configuration of traits such as big head in relation to the body, high prominent forehead, large eyes, chubby cheeks, small mouth, and short rounded limbs—all characteristics of young children or animals—evoke a nurturing, protective response from adults, who perceive such an object as loveable.[28] Toy manufacturers successfully exploit and exaggerate this "infant schema" to make their products cute, cuddly, and appealing.[29] Figures such as "Bambi" and "Snoopy" have been depicted in the same way, and Stephen Jay Gould has traced the progressive juvenilization of Walt Disney's Mickey Mouse over the years as he became more and more popular.[30] In the domestication of animals human beings exerted selection toward neoteny, which refers to the retention of infantile or juvenile traits in the adult form.[31] Results of this process (which can involve behavioral as well as physical traits) are particularly visible in dogs such as Pekinese and Boston terriers, but are also present in other breeds and in almost all domesticated species.[32]

Cartoonist Berryman, consciously or unconsciously, employed this same juvenilizing process in making Roosevelt's bear into an immature, frightened cub with a whimsical expression. The young bear in later cartoons is small, defenseless, and docile, and evokes a sympathetic response. The animal is rendered powerless, subservient to its human captor. Thus the cartoonist set the stage for the creation of the immensely appealing stuffed teddy bear who little resembled its wild antecedent and does so even less as it approaches a century of unflagging popularity. Teddy bears themselves have shown an overall tendency toward increased neoteny over the years since their inception (though with some intermittent exceptions). Older bears generally tend to have more pronounced noses, longer limbs, and more realistic proportions, as compared to later editions which often have larger heads, flatter faces, and shorter, thicker limbs. Whereas older teddy bears were apt to be firmly stuffed, a recent variety is constructed with an extremely soft and yielding body to allow for more effective hugability. As one contemporary observer noted, trends are "moving away from the proud, stiff classical bear toward those that are easier to snuggle, with fluffier fur and bodies. Traditional bears don't mush."[33]

Whenever teddy bear companions are discussed, it is difficult to keep in mind that they are inanimate objects, so powerful are their symbolic attributes. It is commonly stated by owners that each has its own personality, and all manner of thoughts, emotions, and actions are attributed to the bears. These perceptions are not confined to the teddy bear "parents," but are widely shared. When one woman received her degree from a prestigious ivy-league college, she brought her bear to the commencement. When people observed the teddy in a cap and gown, responses such as addressing the bear with the query "what are you going to do after graduation?" were common. This strange, pervasive phenomenon is complex, with deep-lying roots. Teddy bears have diverged greatly from their origins. Although wild cubs are extraordinarily appealing, live bears generally make very unsatisfactory pets, and are trained to do human bidding only by drastic and usually barbarous methods. "Lady Washington," the favorite captive bear of Grizzly Adams by his own admission had been chained to a tree and brutally beaten into submission.[34] At best, the wild bear is not a "buddy," even to its staunchest admirers. Rather, it is a species that requires for its very existence a habitat undisturbed by humankind, and its presence is said to be a "barometer" of the status of the primeval ecology. Bears signify the wilderness and as one grizzly bear researcher noted, they are "the ultimate symbol of nature uncontrolled by people."[35] According to Paul Shepard, "the perception of bears as humanoid undoubtedly has a long history, predating Yogi Bear, Teddy Bear, Smokey the Bear, the Three Bears, and even the Russian Bear, going far back into the bear cults of the last ice age. . . . The great difficulty is that the bear is so easily perceived as a masked man in a fur coat that its true otherness is extremely difficult to see."[36]

Differing so greatly from its antecedents, the teddy bear seems to be the actual bear's counterpart of oppositions, a sort of contrary alter ego to the real bear. Whereas living bears are unpredictable wild creatures whose "hugs" may be injurious or fatal, and who are aloof, ferocious hunters, teddy bears are completely tame, dependent and dependable, caring friends whose hugs combat anxiety and are even curative. To please us, they wear clothes, go to school, read, cook, or ski, wait endless hours for our return, and sleep and eat with us—all without reciprocal demands. Real bears are never subservient. Just as members of the animal kingdom are so like us and yet so different, thereby giving us an image by which to measure our humanity, so the teddy bear, modelled after the living bear and yet so different, gives us a way to reflect upon what has been lost in the transformation. Missing qualities are autonomy, risk, and the wild spirit that cannot be controlled. Teddy bears are neutered, civilized, humanized, and sanitized. Through the process of creating this version of a bear, it seems we have taken their former strength unto ourselves. This power usurped from the animal quiets our anxieties perhaps because it establishes us as total masters. There is no more to fear, because we have reached the limit of conquest. Our charming teddy bears, still not quite dolls, embody the ultimate taming, the extreme polarity in the wild-to-tame transition, the conversion of nature to culture. Dr. Paul C. Horton explains the effects of this wild-tame dichotomy in psychological terms, suggesting that the child achieves a soothing illusion of conflict resolution by controlling the symbolic wild animal. The choice of a representation of a dangerous animal as a soother, he believes, "marks the beginning of the child's ability to seek out and integrate the paradoxical and contradictory."[37]

Living bears are mighty and ferocious, the archetypal predator. It is possible that human beings still carry a vestigial inborn fear of predators, as a survival mechanism inherited from our ancestors who were physically defenseless as compared to the bear, and whose offspring were so vulnerable to predation. By totally subjugating the bear to make it an effigy to cuddle, we have allayed our collective fears, and soothed all manner of anxieties. And yet, there is our characteristic ambivalence. We admire wild nature, even as we fear it and seek to control and destroy it. I believe that the ever-popular teddy bear represents a gigantic collective wish fulfillment of humankind's ancient dream of unity with the animal world. A deep desire to befriend our fellow creatures is attested to by the fact that some members of almost every known society are motivated to keep and cherish pets. Human beings reach out for camaraderie with wild creatures, but are often disappointed by the lack of reciprocity. The teddy bear, however, will not let us down. He mitigates our failure to connect with the animal world at the same time as he softens our hurt and disillusionment with our own kind. He assuages the grief that results from dual alienation from both human and animal society. The strength and intelligence drawn from the real bear and its strikingly human characteristics, with all the antagonistic traits removed in the teddy bear, make it a useful bridge between animal and human spheres. Teddy bears are animals that nearly, but not quite, became people, and serve as buffers from the pains of the human condition.

Appreciation of the qualities that preindustrial peoples admired in bears may still lie deep in the human unconscious, giving the teddy bear some of the living bears' magical virtues that exploitative world views overlook. Because of their winter hibernation and mysterious awakening in the spring, bears represent nature's cycle of renewal and, metaphorically, the hope of immortality. Also drawing from the image of living bears, teddy bears remain symbolic healers with curative knowledge and abilities. They are idealized beings adapted solely to specific human needs, possessing the innocence that comes from their animal roots and the ability to counter loneliness through a projected social perfection. With our teddy bears, whose sinlessness and solicitude sustains us, at least temporarily, we can make our world safer.

Horton, a psychiatrist, has shed light on the teddy bear phenomenon through his studies of transitional relatedness, which is defined as the person's unique experience of an object (called the transitional object) which provides soothing comfort symbolic of the maternal presence. The transitional object, which in our society is most commonly a teddy bear, represents a "spill-over" of attachment to the mother; it is the infant's "first not-me possession" and becomes a vehicle for solace.[38] Horton's research has shown that forty percent of children choose stuffed animals as their first solacer, and of these, sixty percent are teddy bears. He asserts that one child in four derives its first extra-maternal solace from teddy bears. At five to six years of age, sixty-three percent of children use stuffed animals, with sixty percent of these being bears. After that, Horton found that attachment to teddy bears drops off. Then there is a burst of re-attachment in the late teens. According to his studies, eighty percent of female and forty percent of male college freshmen have teddy bears (personal communication).

Psychiatrists have learned that use of transitional objects is nearly universal and is necessary to healthy human development. Horton believes that the achievement of a state of solace is a prerequisite to a person's capacity for positive emotions such as joy and love.[39] He points out that the existence of transitional relatedness shows that a child is already working to master the anxiety that accompanies separation from the mother. One nine-year-old whose parents were divorced described how her stuffed animals helped her deal with being alone after school now that her mother had a job outside their home: "They keep me company. Like when I'm watching TV downstairs, and no one's at home, they watch TV with me. And when it's a funny part, I think they're laughing."[40]

Ever since "Teddy" Roosevelt's refusal to shoot a captive bear was perceived as a gracious gesture toward the natural world, the gentler image of America's most formidable predator has been personified in an immensely appealing and popular toy. The fearsome wild animal, once held to be invulnerable, is now tamed as a teddy bear—a faithful friend, confidante, and protector for vulnerable human beings. For young and old, teddy bears have come to represent trust, love, and reassurance. A billboard on the highway passing through downtown Providence, Rhode Island depicts a teddy bear with a prominent red heart wearing a seat belt. His message is: "Buckle up. Somebody Loves You!" When hurt and neglected by people, the symbolic meaning with which the stuffed bears have been endowed allows them to afford comfort and to personify goodness and kindliness. In a world full of uncertainty about almost everything else, there is no ambivalence about teddy bears: they are unequivocally benign. Their dependability makes them always available when needed, especially for warding off human fears that come with the darkness of night. Teddy bears are nurturers, even as they are nurtured.

As one of society's most cherished shared symbols, the teddy bear denotes the spirit of caring and unconditional affection. Horton says that "teddy bears run deep in American culture because they are a source of requited love for so many."[41] A teddy bear store owner told me that many of her sales involve distraught men who have quarrelled with their wives or girl friends and need a gift to make amends. Although she sells other varieties of stuffed toys, these customers always insist on teddy bears to express their love. Peculiarly adapted to combat the fears and anxieties resulting from the alienation prevalent in modern society, the bear uses its "secret power" not only to provide companionship and solace, but also amusement. It adds gentle humor and fun to many lives. As one adult arctophile who participated in the "All American Teddy Bears' Picnic" noted, "It must sound insane—adults, even grown men with their Teddies in the competition—but we had such a wonderful time."[43] It is not surprising that the protective attributes of teddy bears are utilized in advertisements. Celestial Seasoning's TV commercial for "Sleepy Time Tea," for example, features a bear that, like its product, provides peace and serenity at bedtime. And a recent ad for Kodacolor film that promises to "show your true colors" depicts an apprehensive child clutching a teddy bear during his fist visit to the barber shop, with the caption reading "True Courage." Also available on the market is the epitome of bear protection—a burglar alarm in the form of an "innocent looking teddy bear [that] lets out a mean squawk if jostled by an intruder."[43]

A teddy bear can even quell the anxieties of a future king. According to the author of a recent book about the British royal family, Prince Charles, whose "Old Teddy" has a special place in his heart, sees that the stuffed bear is tucked snugly in bed every night. Teddy "accompanies him everywhere," packed by his valet and taken "with the entourage." In order to safeguard the Prince's image, "the only occasions when teddy is left at home are when Charles visits the regiments where he is colonel in chief."[44] No such exclusion from the military was noted among Americans during the Desert Storm operation. A teddy named "Jeranamo" served as "mascot of the stealth fighters in Saudi Arabia." The eight-inch bear rode "in the cockpit map case of the radar-evading F-117A since the first flight over Baghdad" until the end of the campaign.[45] Newspapers from coast to coast created a great deal of publicity when they featured a front-page photograph of an army nurse, a lieutenant assigned to the 5th Mash Unit in Saudi Arabia, cuddling her teddy bear as she napped on her cot during the Gulf War.[46] The photo, in fact, evoked a sensational response; it was a theme for radio talk shows and countless people called newspapers to find out how to write to or meet the woman. She received numerous marriage proposals. A debate raged about "how this picture of a slumbering blonde with a cuddly bear depicted the roles of women in Desert Storm." But doubts about women at war were dispelled by information that the nurse's assault rifle was almost as close as her bear as she slept and reports that she was highly skilled in Army training exercises and comes from a family with a strong military tradition. Her "family is proud of that photo—teddy bear and all. . . . It shows she's tough enough to be a soldier. And tough enough to be a woman, too."[47]

The image of the teddy bear in contemporary culture is universally positive. Good Bears of the World, an international non-profit organization that provides bears to sick and abused children and the elderly, designated 1985 The Year of the Bear "as a more hopeful follow-up to the doom-filled connotations of George Orwell's 1984." The goal was for "Teddy [to] become a dominant factor in developing fellowship amongst all people, regardless of race, religion, age, or sex."[48] With the aim of making children less aggressive, one group, the Los Angeles Alliance for Survival, exchanges war toys such as guns for teddy bears.[49] During the 1988 presidential campaign, officials who planned Michael Dukakis' strategy for the second debate with George Bush raised the question "Should Dukakis come out fighting, Duke as Rambo? Or should he accentuate the positive, Duke as teddy bear."[50] As events in modern life come more and more to reflect the spirit of Rambo, the teddy bear—epitome of the wild that has been tamed—becomes an even more powerful symbol of the beneficence people seek. According to Paul Shepard, the wild bear "strikes a chord in us of fear and caution, curiosity and fascination. In self-absorption we may fool ourselves, forgetting his otherness, and feed him in a national park as though he were a pet dog." Shepard's observations apply to the teddy bear phenomenon:

> Perhaps the impulse is the same, whether we invite the bear to share our food or our folklore: the urge to be reunited with something lost and treasured, seen in the animal that most resembles us. It is almost as though in him we can see how great is our loss of contact with ourselves. Perhaps the bear can still serve as a pilot and messenger.[51]

Notes

1. Paul Shepard, and Barry Sanders, *The Sacred Paw: The Bear in Nature, Myth, and Literature* (New York: Viking, 1985), xii.
2. Joseph Campbell, *The Way of the Animal Powers* (San Francisco: Harper & Row, 1983), 147, 54–56.
3. W. B. White, *Seeing Stars* (n. p.: Rand McNally & Company, 1942), 7.
4. Irving Hallowel, "Bear Ceremonialism in the Northern Hemisphere," *American Anthropologist,* 28, 1 (1926):1-175.
5. See Frank Speck, *Naskapi: The Savage Hunters of the Labrador Peninsula* (Norman: University of Oklahoma Press, 1977); Richard K. Nelson, *Make Prayers to the Raven: A Koyukon View of the Northern Forest* (Chicago: University of Chicago Press, 1983); and Raymond J. DeMallie, *The Sixth Grandfather: Black Elk's Teachings Given to John Neihardt* (Lincoln: University of Nebraska Press, 1984).
6. Hallowel, "Bear Ceremonialism," 148–49.
7. Nelson, *Raven,* 173.
8. Verner Reed, "The Ute Bear Dance" *American Anthropologist,* 9, 7 (1896):237-248.
9. George Bird Grinnel, *The Cheyenne Indians: Their History and Ways of Life,* 2 vols. (New Haven: Yale University Press, 1923), 105.
10. DeMallie, *Sixth Grandfather,* 178–80.
11. T. P. Vukanovic, "Gypsy Bear Leaders in the Balkan Peninsula," *Journal of the Gypsy Lore Society,* 38, 3-4 (1959):106-127.
12. Elizabeth A. Lawrence, "Relationships With Animals: The Impact of Human Culture," *National Forum,* 66, 1 (Winter, 1986):14-18.
13. See Reed, "Dance," Vukanovic, "Gypsy Bear," and Marion Wood, *Spirits, Heroes & Hunters from North American Indian Mythology* (New York: Schoken Books, 1982), 48–52.
14. See Tom McNamee, "Breath-Holding in Grizzly Country," *Audubon,* 84, 6 (1982):69-83; and Enos Mills, *The Grizzly: Our Greatest Wild Animal* (Boston: Houghton Mifflin, 1919), 9–18.
15. Elizabeth A. Lawrence, *Rodeo: An Anthropologist Looks at the Wild and the Tame* (Chicago: University of Chicago Press, 1982), 251–53.
16. Frederick J. Turner, *The Significance of the Frontier in American History (Annual Report of the American Historical Association 1893),* (Washington, D.C.: Government Printing Office, 1894).
17. Roderick Nash, *Wilderness and the American Mind,* 3d ed (New Haven: Yale

University Press, 1982), 149.

18. Linda Mullins, *Teddy Bears Past & Present* (Cumberland, Maryland: Hobby House Press, 1986), 32–40; Paul Schullery, ed., *American Bears: Selections from the Writings of Theodore Roosevelt* (Boulder: Colorado Associated University Press, 1983), 10.
19. Mullins, *Teddy Bears*, 32–33.
20. Schullery, *American Bears*, 10–11.
21. Jim Seavor, "Most Expensive Teddy," *Providence Journal-Bulletin* (October 4, 1989):E2.
22. Virginia Bohlin, "The Bear Fact of the Matter: Teddy's Not For Just the Young," *Boston Globe* June 1, 1986):108.
23. n.a., "R. I. Uses Teddy Bears for Security on Buses," *Boston Globe* (September 13, 1987).
24. n.a., "Tooth Care Doesn't Have to be a Pain," *USA Today* (September 24, 1986):4D.
25. C. Eugene Emery, Jr. "Teddies Show Tots A Hospital's Heart," *Providence Journal* (September 22, 1986):A1, A2.
26. n.a., "Nature's Clinic: Care Bears Speed Healing," *Woman's World*, 8, 15 (April, 1987):13.
27. Personal communication with the author.
28. Konrad Lorenz, *The Foundations of Ethology* (New York: Simon and Schuster, 1982), 165.
29. Irenaus Eibl-Eibesfeldt, *Love and Hate: The Natural History of Behavior Patterns* (New York: Holt, Rinehart and Winston, 1972), 21–23.
30. Stephen Jay Gould, *The Panda's Thumb* (New York: W. W. Norton & Co., 1980), 95–107.
31. Elizabeth A. Lawrence, "Neoteny in American Perceptions of Animals," *Journal of Psychoanalytic Anthropology*, 9, 1 (Winter, 1986):41-54.
32. See Juliet Clutton-Brock, *Domesticated Animals from Early Times* (Austin: University of Texas Press, 1981).
33. James Hirsch, "Teddy Bear's Work Is Never Done," *New York Times* (September 28, 1988):C1, C2.
34. Theodore Hittell, "Lady Washington," in *The Grizzly Bear*, Bessie Doak Haynes and Edgar Haynes, eds. (Norman: University of Oklahoma Press, 1967), 102–03.
35. Amos Enos, "Symbol of Wild America," *Audubon Action*, 1, 2 (February, 1983):11.
36. Paul Shepard, *Thinking Animals: Animals and the Development of Human*

Intelligence (New York: Viking, 1978), 229.

37. Paul C. Horton, *Solace: The Missing Dimension in Psychiatry* (Chicago: University of Chicago Press, 1981), 107, 115.
38. Horton, *Solace*, 18, 35; Paul C. Horton, Herbert Gewirtz, and Karole J. Kreutter, eds., *The Solace Paradigm: An Eclectic Search for Psychological Immunity* (Madison, Conn.: International Universities Press, 1988), 323.
39. See Horton et al., *Solace Paradigm.*
40. Horton, *Solace*, 160; Horton et al., *Solace Paradigm*, 260.
41. Hirsch, "Teddy Bear's Work."
42. Shirley Utterback, "Bristol Women Win 'Teddy' Event," *Providence Journal-Bulletin* (July 30, 1987):3.
43. n.a., "Innocent Looking Teddy Bear," *The Providence Journal* (September 3, 1990):D1.
44. Andrew Morton, "Diana's Diary," *Good Housekeeping*, 211, 1 (1990):112-115; 158-162.
45. n.a., "Mascot," *USA Today* (January 21, 1991):5.
46. n.a., "On the Front with Desert Storm," *USA Today* (February 25, 1991):SA.
47. J. Taylor Buckley, "The Story Behind the Teddy Bear Photo," *USA Today* (March 1, 1991):11A.
48. Mullins, *Teddy Bears*, 295.
49. Melanie DuLac, "California Group Exchanges Toy Guns for Teddy Bears," *Sunday Independent South* (January 8, 1989):7.
50. Richard Stengel, "The Election: Nine Key Moments," *Time*, 132, 21 (November 21, 1988):48-56.
51. Shepard, *Sacred Paw*, xii.

Appendix I

PAUL SHEPARD'S BIBLIOGRAPHY

BOOKS:

Current project. *The Eclectic Primitive: Nature and History,* (cf., "A Post-Historic Primitivism" in Articles).

Forthcoming: *The Others: How Animals Made Us Human,* Shearwater Books.

Man in the Landscape: An Historic View of the Esthetics of Nature, Texas A & M University Press, College Station, TX, 1991.
Ibid. Ballantine, New York, 1972.
Ibid. Knopf, New York, 1967.

The Sacred Paw: The Bear in Nature, Myth and Literature (with Barry Sanders), Arcana Books, Penguin, New York, 1992.
Ibid., The Viking Press, New York, 1985.

Nature and Madness, Sierra Club Books, San Francisco, 1982.

Thinking Animals: Animals and the Development of Human Intelligence, The Viking Press, New York, 1978.

The Tender Carnivore and the Sacred Game, Scribners, New York, 1973.

Environ/mental: Essays on the Planes as a Home, (with Daniel McKinley) Houghton Mifflin, Boston, 1971.

The Subversive Science: Essays Toward an Ecology of Man, (with Daniel McKinley), Houghton Mifflin, Boston, 1969.

The Pictorial History of the 493 Armored Field Artillery Battalion, 1943-1946, E. Kieser, Augsburg, 1945.

MONOGRAPHS:

English Attitudes Towards the New Zealand Landscape Before 1850, Pacific Viewpoint Monographs #4, Wellington, 1967.

ARTICLES:

"A Post-Historic Primitivism," in Max Oelschlaeger, ed., *The Wilderness Condition: Essays on Environment and Civilization,* Sierra Club Books, San Francisco, 1992.

"Seeking Out Kindred Spirits," *Parabola* issue on "Hunting," Summer, 1991.

"Digging for Our Roots," *Places,* Fall, 1990.

"The Garden as Objet Trouve," in Randolf Hester and eds., *The Meanings of Gardens,* Johns Hopkins Univ. Press, Baltimore, 1988.

"Introduction," Japanese edition, Edith Cobb, *The Ecology of Imagination in Childhood,* Shisaku-sha Publ., Tokyo, 1986.

"Celebrations of the Bear," (with Barry Sanders), *North American Review,* Sept. 1985.

"The Philosopher, the Naturalist and the Agony of the Planet," in Randall Eaton, ed., *The Human-Animal Bond,* Sierra Nevada College Press, Incline Village, 1985.

"Introduction," Ortega y Gassett, *Meditations on Hunting,* Scribners, New York, 1985, 1974.

"If You Care About Nature You Can't Go on Hating the Germans Like This," in Michael Tobias, ed., *Humanity and Radical Will,* Avant Books, Los Angeles, 1984.

"The Ark of the Mind," *Parabola,* Spring, 1983.

"Five Green Thoughts," *Massachusetts Review;* Summer, 1980.

"Introduction," *Joseph Meeker: The Comedy of Survival,* Guild of Tutors Edition, Los Angeles, 1980.

"Ecosophy," in J. Sherwood Weber, ed., *Good Reading,* Waybright and Talley, New York, 1979.

"Place in American Culture," *North American Review,* Fall, 1977.

"Place and Human Development," Proceedings, *Symposium on Children and Urban Environment,* U.S. Forest Service and the Gifford Pinchot Institute, Washington, D.C., 1977.

"Animal Rights and Human Rites", *North American Review;* Winter, 1975.

La Ecologia y el hombre," *Revista de Occidente* #143 y 144, 1976.

"Whatever Happened to Human Ecology?," *BioScience,* Dec., 1967.

"The Virtues of Anonymity," *The Saturday Review,* Sept. 17, 1966, repr. in Martha Healey Cox, ed., *A Reading Approach to College Writing,* Chandler Pub. Co., 1967, and in Keith Huntress, Design for Reading, American Book Co., 1968.

"Biological Sciences," in J. Sherwood Weber, *Good Reading,* Mentor, N.Y., 1964.

"The Corvidean Millennium, or Letter From an Old Crow," *Perspectives in Biology and Medicine,* Spring, 1964.

"The Arboreal Eye," *School Science and Mathematics,* Dec., 1964.

"The Tender Carnivore," *Landscape* 14 (1), 1964.

"The Artist as Explorer," *Landscape* 12 (2), 1962.

"The Cross Valley Syndrome," *Landscape* 10 (3), 1960.

"A Theory of the Value of Hunting," Transactions, *24th North American Wildlife Conference,* Washington, D.C., 1959.

"Reverence for Life at Lambarene," *Landscape* 8 (2), 1958.

"The Wilderness as Nature," *Atlantic Naturalist,* April, 1958.

"Paintings of the New England Landscape," *College Art Journal,* Fall, 1957.

"Dead Cities in the American West," *Landscape* 6 (2), 1956.

"The Nature of Tourism," *Landscape* 5 (1), 1955.

"Montana's Marching Mountains," *Nature Magazine*, Feb., 1954.

"They Painted What They Saw," *Landscape* 3 (1), 1953.

"Something Amiss in the National Parks," *Yale Conservation Studies*, Vol. II, Yale Conservation Program, New Haven, 1953.

"The Outdoor Public—an Unexploited Educational Resource," *Yale Conservation Studies*, Vol. I, Yale Conservation Program, New Haven, 1952, repr. in Bulletin de l'Association pour la Defense de l'Ourthe et de ses affluents, Paris, July, 1953.

"Experiment in Learning," *American Forests*, Oct. 1952.

"Can the Cahow Survive?" *Natural History*, July, 1952, repr. in Edward M. Weyer, ed., *The Illustrated Library of Natural Sciences*, Vol. I, Simon & Schuster, N.Y., 1958.

"The Dove is Doubtful Game," *Nature Magazine*, Sept., 1952.

"Our Highways and Wildlife," *Nature Magazine*, Jan. 1952.

"Eyes, Clues to Life Habits," *Nature Magazine*, Summer, 1951, repr. in Richard W. Westwood, ed., *This is Nature, Thirty Years of the Best from Nature Magazine*, N.Y., 1952.

EDITOR'S NOTE: This bibliography is neither exhaustive nor definitive; it does list those works singled out by Paul Shepard for inclusion in this bibliography.

Notes on the Contributors

Morris Berman is a social critic and cultural historian. He held the Garrey Carruthers Chair in Honors at the University of New Mexico, 1994-95. His books include *Social Change and Scientific Organization, The Reenchantment of the World,* and *Coming to Our Senses: Body and Spirit in the Hidden History of the West.*

Daniel Sloan Brown is a student and teacher of ecopsychology. He is also founder of New Earth Nutritionals, an ecologically-sane business. He is currently working on a dissertation exploring the psychodynamics and phenomenology of ecological consciousness.

David Clarke Burks is a writer and teacher in the Environmental Studies Program at the University of Oregon. He is editor of the anthology *Place of the Wild.*

J. Baird Callicott is a Professor of Philosophy and Natural Resources at the University of Wisconsin–Stevens Point. His books include *Earth's Insights: A Multicultural Survey of Ecological Ethics from the Mediterranean Basin to the Australian Outback, In Defense of the Land Ethic,* and *A Companion to A Sand County Almanac.*

Shagbark Hickory calls the prairies of Southeastern Wisconsin and the mountains and deserts of Idaho his homes. His many articles include "Postmodern Environmental Ethics: Ethics as Bioregional Narrative" and "Eco-Feminism and Deep Ecology."

John Cobb, Jr., is an ecotheologian. His many books include *Is It Too Late? A Theology of Ecology* and *For the Common Good,* with Herman Daly.

Michael P. Cohen is a professor at Southern Utah University. His books include *The Pathless Way* and *The History of the Sierra Club.*

Paul Faulstich is Assistant Professor of Environmental Studies at Pitzer College. Previously, he taught anthropology at Linfield College, served as Curator of Native American Art at the Portland Art Museum, and was a Degree Associate with the East-West Center in Honolulu. As an interdisciplinary human ecologist, his academic interests range from traditional ecological knowledge to the ecology of personhood. He makes his home on the alluvium of the Thomson Creek watershed, Claremont, California.

Chellis Glendinning is a psychologist focusing on the interplay of psychology, society, technology, and the natural world. Her books include *My Name Is Chellis, I'm In Recovery From Western Civilization* and *When Technology Wounds: The Human Consequences of Progress.*

Carl H. Hertel is a Professor of Art and Environmental Studies at Pitzer College, Claremont, California.

Robert Hood is a doctoral student teaching in the Department of Philosophy at Bowling Green State University. His writings include "Discursive Horizons of Human Identity" and "Wilderness in Postmodern Environmental Ethics."

Florence Krall is a teacher and nature writer. She is the author of *Ecotone: Wayfaring on the Margins.*

Kraig Klungness is a wilderness advocate and student of deep ecology. He lives near the shore of Lake Superior in the Upper Peninsula of Michigan where he works for the preservation of wildlands and writes about ecological issues. He would like to thank Kate Alvord, Thomas Lyon, and Stephanie Mills for their helpful comments in reviewing this essay.

Dolores LaChapelle is director of the Way of the Mountain Learning Center in the San Juan Mountains of Colorado. Her books include *Earth Festivals, Earth Wisdom, Sacred Land, Sacred Sex: Rapture of the Deep,* and *Deep Powder Snow: 40 Years of Ecstatic Skiing, Avalanches, and Earth Wisdom.*

Elizabeth Atwood Lawrence, a veterinarian as well as a cultural anthropologist, is a Professor of Environmental Studies at Tufts University School of Veterinary Medicine, where she teaches and carries out research in the field of human-animal relationships. She is the author of *Rodeo: An Anthropologist Looks at the Wild and the Tame, Hoofbeats and Society: Studies of Human-Horse Interactions, His Very Silence Speaks: Comanche—The Horse Who Survived Custer's Last Stand,* the text for *The Horse: Photographic Images 1839 to the Present,* and numerous book chapters and journal articles.

John A. Livingston is a teacher and wilderness advocate. His books include *The Fallacy of Wildlife Conservation* and *Rogue Primate.*

Michael Martin McCarthy is a Professor at Texas A&M University, where he was formerly Dean of the College of Architecture. Prior to that he held the Murdoch Professorship of Landscape Architecture and Environmental Planning at the University of Melbourne in Australia.

Joseph Meeker is a teacher and writer. His books include *The Comedy of Survival* and *Minding The Earth.*

Ralph Metzner is a psychotherapist and Professor of Psychology at the California Institute of Integral Studies. His books include *Maps of Consciousness* and *The Well of Remembrance.*

James A. Mischke is Chair of Social and Behavioral Sciences Division of Navajo Community College-Shiprock (New Mexico). He welcomes correspondence on his essay to his home address: 143 E. North St., Cortez, CO 81321.

Max Oelschlaeger is a Professor of Philosophy at the University of North Texas. His books include *The Idea of Wilderness* and *Caring for Creation.*

William A. Reiners is Professor of Botany at the University of Wyoming.

Charles Schweighauser is a Professor of Astronomy/Physics, English, and Environmental Studies at the University of Illinois at Springfield. His books include *Astronomy from A to Z.*

George Sessions is a Professor of Philosophy at Sierra College. His books include (with Bill Devall) *Deep Ecology: Living as if Nature Mattered* and *Deep Ecology for the 21st Century.*

Gary Snyder is a poet and essayist. His books include *Turtle Island* (awarded the Pulitzer prize for Poetry in 1975) and *The Practice of the Wild.*

Laura Westra is a Professor at University of Windsor (Canada). She is the author of *An Environmental Proposal for Ethics.*

Douglas L. Wilson was a colleague of Paul Shepard at Knox College and is now Director of the International Center for Jefferson Studies at Monticello, Charlottesville, Virginia. He is the editor of *Jefferson's Literary Commonplace Book* and *The Genteel Tradition: Nine Essays by George Santayana.*

KIVAKÍ PRESS

Kivakí Press publishes non-fiction books that provide access to an ecological experience of life and place.

If you are interested in receiving a free catalog of current titles, please contact us—

by mail:

Kivakí Press
585 East 31st Street
Durango, CO 81301 USA

by phone/fax:

(800) 578-5904 (within the U.S.A.)
(970) 385-1767
(970) 385-1974

by e-mail (Internet):

kivaki@frontier.net